Student Solutions Manual
Volume 3
for
Tipler and Mosca's
Physics for Scientists and Engineers
Sixth Edition

David Mills
Professor Emeritus
College of the Redwoods

W. H. Freeman and Company
New York

© 2008 by W. H. Freeman and Company

ISBN-13: 978-1-4292-0301-2 (Volume 3: Chapters 34–41)
ISBN-10: 1-4292-0301-3

Printed in the United States of America

First printing

W. H. Freeman and Company
41 Madison Avenue
New York, NY 10010
Houndmills, Basingstoke
RG21 6XS, England
www.whfreeman.com

Contents

Acknowledgments

Gene Mosca (formerly of the United States Naval Academy and co-author of the Sixth Edition) helped clarify and otherwise improve many of my solutions and provided guidance when I was unsure how best to proceed. It was a pleasure to collaborate with Gene in the creation of this solutions manual. He shares my hope that you will find these solutions useful in learning physics.

Carlos Delgado (Community College of Southern Nevada) and Mike Crivello (San Diego Mesa College) checked the solutions. Without their thorough work, many errors would have remained to be discovered by the users of this solutions manual. Carlos also suggested several alternate solutions, all of which were improvements on mine, and they are included in the solutions manual. Their assistance is greatly appreciated. In spite of the best efforts of Carlos and Mike, there may still be errors in some of the solutions, and for those I assume full responsibility. Should you find errors or think of alternative solutions that you would like to call to my attention, please do not hesitate to send them to me by using asktipler@whfreeman.com.

It was a pleasure to work with Susan Brennan, Clancy Marshall, and Kharissia Pettus who guided us through the creation of this solutions manual. I would also like to thank Kathryn Treadway and Janie Chan for organizing the reviewing and error-checking process.

June 2007

David Mills
Professor Emeritus
College of the Redwoods

To the Student

This solution manual accompanies *Physics for Scientists and Engineers,* 6e, by Paul Tipler and Gene Mosca. Following the structure of the solutions to the worked Examples in the text, we begin a solution to an end-of-chapter numerical problem by picturing the problem—representing the problem pictorially whenever appropriate, and expressing the physics of the solution in the form of a mathematical model. Then, the problem is solved or any intermediate steps are filled in as needed, the appropriate substitutions and algebraic simplifications are made, and the solution with the substitution of numerical values (including their units) is completed. This problem-solving strategy is used by experienced learners of physics, and it is our hope that you will see the value in such an approach to problem solving and learn to use it consistently.

Believing that it will maximize your learning of physics, we encourage you to create your own solution before referring to the solutions in this manual. You may find that, by following this approach, you will find different, but equally valid, solutions to some of the problems. In any event, studying the solutions contained herein without having first attempted the problems will do little to help you learn physics.

Chapter 34
Wave-Particle Duality and Quantum Physics

Conceptual Problems

1 • The quantized character of electromagnetic radiation is observed by (*a*) the Young double-slit experiment, (*b*) diffraction of light by a small aperture, (*c*) the photoelectric effect, (*d*) the J. J. Thomson cathode-ray experiment.

Determine the Concept The Young double-slit experiment and the diffraction of light by a small aperture demonstrated the wave nature of electromagnetic radiation. J. J. Thomson's experiment showed that the rays of a cathode-ray tube were deflected by electric and magnetic fields and therefore must consist of electrically charged particles. Only the photoelectric effect requires an explanation based on the quantization of electromagnetic radiation. $\boxed{(c)}$ is correct.

3 • The work function of a surface is ϕ. The threshold wavelength for emission of photoelectrons from the surface is equal to (*a*) hc/ϕ, (*b*) ϕ/hf, (*c*) hf/ϕ, (*d*) none of above.

Determine the Concept The work function is equal to the minimum energy required to remove an electron from the material. A photon that has that energy also has the threshold wavelength required for photoemission. Thus, $hf = \phi$. In addition, $c = f\lambda$. It follows that $hc/\lambda_t = \phi$, so $\lambda_t = hc/\phi$ and $\boxed{(a)}$ is correct.

Estimation and Approximation

11 •• [SSM] During an advanced physics lab, students use X rays to measure the Compton wavelength, λ_C. The students obtain the following wavelength shifts $\lambda_2 - \lambda_1$ as a function of scattering angle θ:

θ	45°	75°	90°	135°	180°
$\lambda_2 - \lambda_1$	0.647 pm	1.67 pm	2.45 pm	3.98 pm	4.95 pm

Use their data to estimate the value for the Compton wavelength. Compare this number with the accepted value.

Picture the Problem From the Compton-scattering equation we have $\lambda_2 - \lambda_1 = \lambda_C(1 - \cos\theta)$, where $\lambda_C = h/m_e c$ is the Compton wavelength. Note that this equation is of the form $y = mx + b$ provided we let $y = \lambda_2 - \lambda_1$ and

$x = 1 - \cos\theta$. Thus, we can linearize the Compton equation by plotting $\Delta\lambda = \lambda_2 - \lambda_1$ as a function of $1 - \cos\theta$. The slope of the resulting graph will yield an experimental value for the Compton wavelength.

(*a*) The spreadsheet solution is shown below. The formulas used to calculate the quantities in the columns are as follows:

Cell	Formula/Content	Algebraic Form
A3	45	θ (deg)
B3	$1 - \cos(A3*PI()/180)$	$1 - \cos\theta$
C3	6.47E^−13	$\Delta\lambda = \lambda_2 - \lambda_1$

θ	$1 - \cos\theta$	$\lambda_2 - \lambda_1$
(deg)		
45	0.293	6.47E−13
75	0.741	1.67E−12
90	1.000	2.45E−12
135	1.707	3.98E−12
180	2.000	4.95E−12

The following graph was plotted from the data shown in the above table. Excel's "Add Trendline" was used to fit a linear function to the data and to determine the regression constants.

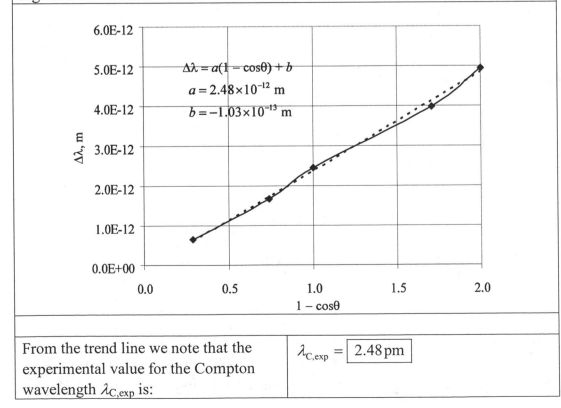

From the trend line we note that the experimental value for the Compton wavelength $\lambda_{C,exp}$ is:

$\lambda_{C,exp} = \boxed{2.48\,pm}$

The Compton wavelength is given by:	$$\lambda_C = \frac{h}{m_e c} = \frac{hc}{m_e c^2}$$
Substitute numerical values and evaluate λ_C:	$$\lambda_C = \frac{1240\,\text{eV} \cdot \text{nm}}{5.11 \times 10^5\,\text{eV}} = 2.43\,\text{pm}$$
Express the percent difference between λ_C and $\lambda_{C,\text{exp}}$:	$$\%\,\text{diff} = \frac{\lambda_{C,\text{exp}} - \lambda_{\text{exp}}}{\lambda_{\text{exp}}} = \frac{\lambda_{C,\text{exp}}}{\lambda_{\text{exp}}} - 1$$ $$= \frac{2.48\,\text{pm}}{2.43\,\text{pm}} - 1 \approx \boxed{2\%}$$

The Particle Nature of Light: Photons

19 • Lasers used in a telecommunications network typically produce light that has a wavelength near 1.55 μm. How many photons per second are being transmitted if such a laser has an output power of 2.50 mW?

Picture the Problem The number of photons per unit volume is, in turn, the ratio of the power of the laser to the energy of the photons and the volume occupied by the photons emitted in one second is the product of the cross-sectional area of the beam and the speed at which the photons travel; i.e., the speed of light.

Relate the number of photons emitted per second to the power of the laser and the energy of the photons:

$$N = \frac{P}{E} = \frac{P\lambda}{hc}$$

Substitute numerical values and evaluate N:

$$N = \frac{(2.50\,\text{mW})(1.55\,\mu\text{m})}{(6.626 \times 10^{-34}\,\text{J} \cdot \text{s})(2.998 \times 10^8\,\text{m/s})}$$

$$= \boxed{1.95 \times 10^{16}\,\text{s}^{-1}}$$

Electrons and Matter Waves

39 • An electron microscope uses electrons that have energies equal to 70 keV. Find the wavelength of these electrons.

Picture the Problem We can approximate the wavelength of 70-keV electrons using $\lambda = \frac{1.226}{\sqrt{K}}\,\text{nm}$, where K is in eV. This solution is, however, only

approximately correct because at the given energy the speed of the electron is a significant fraction of the speed of light. The solution presented is valid only in the non-relativistic limit $v \ll c$.

Relate the wavelength of the electrons to their kinetic energy:	$\lambda = \dfrac{1.23}{\sqrt{K}}\,\text{nm}$
Substitute numerical values and evaluate λ:	$\lambda = \dfrac{1.226}{\sqrt{70 \times 10^3\,\text{eV}}}\,\text{nm} = \boxed{4.6\,\text{pm}}$

General Problems

55 •   Photons in a uniform 4.00-cm-diameter light beam have wavelengths equal to 400 nm and the beam has an intensity of 100 W/m². (*a*) What is the energy of each photon in the beam? (*b*) How much energy strikes an area of 1.00 cm² perpendicular to the beam in 1 s? (*c*) How many photons strike this area in 1.00 s?

Picture the Problem We can use the Einstein equation for photon energy to find the energy of each photon in the beam. The intensity of the energy incident on the surface is the ratio of the power delivered by the beam to its delivery time. Hence, we can express the energy incident on the surface in terms of the intensity of the beam.

(*a*) Use the Einstein equation for photon energy to express the energy of each photon in the beam:	$E_{photon} = hf = \dfrac{hc}{\lambda}$
Substitute numerical values and evaluate E_{photon}:	$E_{photon} = \dfrac{1240\,\text{eV} \cdot \text{nm}}{400\,\text{nm}} = 3.100\,\text{eV}$ $= \boxed{3.10\,\text{eV}}$
(*b*) Relate the energy incident on a surface of area A to the intensity of the beam:	$E = IA\Delta t$
Substitute numerical values and evaluate E:	$E = (100\,\text{W/m}^2)(1.00 \times 10^{-4}\,\text{m}^2)(1.00\,\text{s})$ $= 0.0100\,\text{J} \times \dfrac{1\,\text{eV}}{1.602 \times 10^{-19}\,\text{J}}$ $= 6.242 \times 10^{16}\,\text{eV} = \boxed{6.24 \times 10^{16}\,\text{eV}}$

(c) Express the number of photons striking this area in 1.00 s as the ratio of the total energy incident on the surface to the energy delivered by each photon:

$$N = \frac{E}{E_{photon}} = \frac{6.242 \times 10^{16} \text{ ev}}{3.100 \text{ eV}}$$

$$= \boxed{2.08 \times 10^{16}}$$

61 •• A 100-W source radiates light of wavelength 600 nm uniformly in all directions. An eye that has been adapted to the dark has a 7-mm-diameter pupil and can detect the light if at least 20 photons per second enter the pupil. How far from the source can the light be detected under these rather extreme conditions?

Picture the Problem We can relate the fraction of the photons entering the eye to ratio of the area of the pupil to the area of a sphere of radius R. We can find the number of photons emitted by the source from the rate at which it emits and the energy of each photon which we can find using the Einstein equation.

Letting r be the radius of the pupil, $N_{entering\ eye}$ the number of photons per second entering the eye, and $N_{emitted}$ the number of photons emitted by the source per second, express the fraction of the light energy entering the eye at a distance R from the source:

$$\frac{N_{entering\ eye}}{N_{emitted}} = \frac{A_{eye}}{4\pi R^2}$$

$$= \frac{\pi r^2}{4\pi R^2}$$

$$= \frac{r^2}{4R^2}$$

Solving for R yields:

$$R = \frac{r}{2}\sqrt{\frac{N_{emitted}}{N_{entering\ eye}}} \qquad (1)$$

Find the number of photons emitted by the source per second:

$$N_{emitted} = \frac{P}{E_{photon}}$$

Using the Einstein equation, express the energy of the photons:

$$E_{photon} = \frac{hc}{\lambda}$$

Substitute numerical values and evaluate E_{photon}:

$$E_{photon} = \frac{1240 \text{ eV} \cdot \text{nm}}{600 \text{ nm}} = 2.07 \text{ eV}$$

Substitute and evaluate $N_{emitted}$:

$$N_{emitted} = \frac{100 \text{ W}}{(2.07 \text{ eV})\left(1.602 \times 10^{-19} \dfrac{\text{J}}{\text{eV}}\right)}$$

$$= 3.02 \times 10^{20} \text{ s}^{-1}$$

Substitute for N_{emitted} in equation (1) and evaluate R:

$$R = \frac{3.5\,\text{mm}}{2}\sqrt{\frac{3.02\times10^{20}\,\text{s}^{-1}}{20\,\text{s}^{-1}}}$$

$$\approx \boxed{7\times10^3\,\text{km}}$$

67 ••   The Pauli exclusion principle states that no more than one electron may occupy a particular quantum state at a time. Electrons intrinsically occupy two spin states. Therefore, if we wish to model an atom as a collection of electrons trapped in a one-dimensional box, no more than two electrons in the box can have the same value of the quantum number n. Calculate the energy that the most energetic electron(s) would have for the uranium atom that has an atomic number 92. Assume the box has a length of 0.050 nm and the electrons are in the lowest possible energy states. How does this energy compare to the rest energy of the electron?

Picture the Problem We can use the expression for the energy of a particle in a well to find the energy of the most energetic electron in the uranium atom.

Relate the energy of an electron in the uranium atom to its quantum number n:

$$E_n = n^2\left(\frac{h^2}{8m_e L^2}\right)$$

Substitute numerical values and evaluate E_{92}:

$$E_{92} = (92)^2\left[\frac{\left(6.626\times10^{-34}\,\text{J}\cdot\text{s}\right)^2}{8\left(9.109\times10^{-31}\,\text{kg}\right)\left(0.050\,\text{nm}\right)^2}\times\frac{1\,\text{eV}}{1.602\times10^{-19}\,\text{J}}\right]=1.273\,\text{MeV}$$

$$= \boxed{1.3\,\text{MeV}}$$

The rest energy of an electron is:

$$m_e c^2 = \left(9.109\times10^{-31}\,\text{kg}\right)\left(2.998\times10^8\,\text{m/s}\right)^2\left(\frac{1\,\text{eV}}{1.602\times10^{-19}\,\text{J}}\right)=0.512\,\text{MeV}$$

Express the ratio of E_{92} to $m_e c^2$:

$$\frac{E_{92}}{m_e c^2} = \frac{1.273\,\text{MeV}}{0.512\,\text{MeV}} \approx 2.5$$

The energy of the most energetic electron is approximately 2.5 times the rest-energy of an electron.

71 ••   (*a*) Show that for large n, the fractional difference in energy between state n and state $n + 1$ for a particle in a one-dimensional box is given approximately by $\left(E_{n+1} - E_n\right)/E_n \approx 2/n$ (b) What is the approximate percentage energy difference between the states $n_1 = 1000$ and $n_2 = 1001$? (c) Comment on how this result is related to Bohr's correspondence principle.

Picture the Problem We can use the fact that the energy of the nth state is related to the energy of the ground state according to $E_n = n^2 E_1$ to express the fractional change in energy in terms of n and then examine this ratio as n grows without bound.

(a) Express the ratio $(E_{n+1} - E_n)/E_n$:

$$\frac{E_{n+1} - E_n}{E_n} = \frac{(n+1)^2 - n^2}{n^2} = \frac{2n+1}{n^2}$$

$$= \frac{2}{n} + \frac{1}{n^2} \approx \boxed{\frac{2}{n}}$$

for $n \gg 1$.

(b) Evaluate $\dfrac{E_{1001} - E_{1000}}{E_{1000}}$:

$$\frac{E_{1001} - E_{1000}}{E_{1000}} \approx \frac{2}{1000} = \boxed{0.2\%}$$

(c) Classically, the energy is continuous. For very large values of n, the energy difference between adjacent levels is infinitesimal.

Chapter 35
Applications of the Schrödinger Equation

The Harmonic Oscillator

5 •• Use the procedure of Example 35-1 to verify that the energy of the first excited state of the harmonic oscillator is $E_1 = \frac{3}{2}\hbar\omega_0$. (Note: *Rather than solve for a again, use the step-8 result* $a = \frac{1}{2}m\omega_0/\hbar$ *obtained in Example 35-1.*)

Picture the Problem We can differentiate $\psi(x)$ twice and substitute in the Schrödinger equation for the harmonic oscillator. Substitution of the given value for a will lead us to an expression for E_1.

The wave function for the first excited state of the harmonic oscillator is:

$$\psi_1(x) = A_1 x e^{-ax^2}$$

Compute $d\psi_1(x)/dx$:

$$\frac{d\psi_1(x)}{dx} = \frac{d}{dx}\left[A_1 x e^{-ax^2}\right] = A_1 e^{-ax^2}$$

Compute $d^2\psi_1(x)/dx^2$:

$$\frac{d^2\psi_1(x)}{dx^2} = \frac{d}{dx}\left[A_1 e^{-ax^2}\right] = -2axA_1 e^{-ax^2} - 4axA_1 e^{-ax^2} + 4a^2 x^3 A_1 e^{-ax^2}$$
$$= \left(4a^2 x^3 - 6ax\right)A_1 e^{-ax^2}$$

Substitute in the Schrödinger equation to obtain:

$$-\frac{\hbar^2}{2m}\left[\left(4a^2 x^3 - 6ax\right)A_1 e^{-ax^2}\right] + \frac{1}{2}m\omega_0^2 x^2 A_1 x e^{-ax^2} = E_1 A_1 x e^{-ax^2}$$

Dividing out $A_1 e^{-ax^2}$ (one can do this because the exponential function is never zero) yields:

$$-\frac{\hbar^2}{2m}\left[\left(4a^2 x^3 - 6ax\right)\right] + \frac{1}{2}m\omega_0^2 x^3 = E_1 x$$

or

$$-\frac{\hbar^2}{2m}\left(4a^2 x^3\right) + \frac{\hbar^2}{2m}\left(6ax\right) + \frac{1}{2}m\omega_0^2 x^3 = E_1 x$$

Substitute for a to obtain:

$$-\frac{\hbar^2}{2m}4\left(\frac{m\omega_0}{2\hbar}\right)^2 x^3 + \frac{\hbar^2}{2m}6\left(\frac{m\omega_0}{2\hbar}\right)x + \tfrac{1}{2}m\omega_0^2 x^3 = E_1 x$$

Solve for E_1 to obtain:

$$E_1 = \boxed{\tfrac{3}{2}\hbar\omega_0} = 3E_0$$

Reflection and Transmission of Electron Waves: Barrier Penetration

13 ••   A particle that has mass m is traveling in the direction of increasing x. The potential energy of the particle is equal to zero everywhere in the region $x < 0$ and is equal to U_0 everywhere in the region $x > 0$, where $U_0 > 0$. (*a*) Show that if the total energy is $E = \alpha U_0$, where $\alpha \geq 1$, then the wave number k_2 in the region $x > 0$ is given by $k_2 = k_1\sqrt{(\alpha - 1)/\alpha}$, where k_1 is the wave number in the region $x < 0$. (*b*) Using a **spreadsheet** program or graphing calculator, graph the reflection coefficient R and the transmission coefficient T as functions of α, for $1 \leq \alpha \leq 5$.

Picture the Problem We can use the total energy of the particle in the region $x > 0$ to express k_2 in terms of α and k_1. Knowing k_2 in terms of k_1, we can use $R = \dfrac{(k_1 - k_2)^2}{(k_1 + k_2)^2}$ to find R and $T = 1 - R$ to determine the transmission coefficient T.

(*a*) Using conservation of energy, express the energy of the particle in the region $x > 0$:

$$\frac{\hbar^2 k_2^2}{2m} + U_0 = \alpha U_0$$

Solving for k_2 gives:

$$k_2 = \frac{\sqrt{2mU_0(\alpha - 1)}}{\hbar}$$

From the equation for the total energy of the particle:

$$k_1 = \frac{\sqrt{2m\alpha U_0}}{\hbar}$$

Express the ratio of k_2 to k_1:

$$\frac{k_2}{k_1} = \frac{\dfrac{\sqrt{2mU_0(\alpha-1)}}{\hbar}}{\dfrac{\sqrt{2m\alpha U_0}}{\hbar}} = \sqrt{\frac{\alpha-1}{\alpha}}$$

and $\boxed{k_2 = \sqrt{\dfrac{\alpha-1}{\alpha}}\,k_1}$

(b) The reflection coefficient R is given by:

$$R = \frac{(k_1 - k_2)^2}{(k_1 + k_2)^2}$$

Factor k_1 from the numerator and denominator to obtain:

$$R = \frac{\left(1 - \dfrac{k_2}{k_1}\right)^2}{\left(1 + \dfrac{k_2}{k_1}\right)^2}$$

Substitute the result from Part (a) for k_2/k_1:

$$R = \frac{\left(1 - \sqrt{\dfrac{\alpha-1}{\alpha}}\right)^2}{\left(1 + \sqrt{\dfrac{\alpha-1}{\alpha}}\right)^2} = \left(\frac{1 - \sqrt{\dfrac{\alpha-1}{\alpha}}}{1 + \sqrt{\dfrac{\alpha-1}{\alpha}}}\right)^2$$

The transmission coefficient is given by:

$$T = 1 - R = 1 - \left(\frac{1 - \sqrt{\dfrac{\alpha-1}{\alpha}}}{1 + \sqrt{\dfrac{\alpha-1}{\alpha}}}\right)^2$$

A spreadsheet program to plot R and T as functions of α is shown below. The formulas used to calculate the quantities in the columns are as follows:

Cell	Content/Formula	Algebraic Form
A2	1.0	α
B2	(1−SQRT((A2−1)/A2))/ (1+SQRT((A2−1)/A2))^2	$\left(\dfrac{1 - \sqrt{\dfrac{\alpha-1}{\alpha}}}{1 + \sqrt{\dfrac{\alpha-1}{\alpha}}}\right)^2$

C2	1–B2	$1 - \left(\dfrac{1 - \sqrt{\dfrac{\alpha - 1}{\alpha}}}{1 + \sqrt{\dfrac{\alpha - 1}{\alpha}}} \right)^2$

	A	B	C
1	α	R	T
2	1.0	1.000	0.000
3	1.2	0.298	0.702
4	1.4	0.198	0.802
5	1.6	0.149	0.851
18	4.2	0.036	0.964
19	4.4	0.034	0.966
20	4.6	0.032	0.968
21	4.8	0.031	0.969
22	5.0	0.029	0.971

The following graph was plotted using the data in the above table:

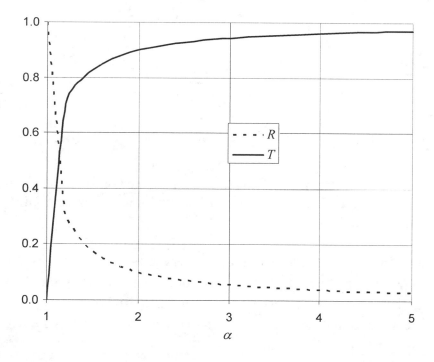

15 •• A 10-eV electron (an electron with a kinetic energy of 10 eV) is incident on a potential-energy barrier that has a height equal to 25 eV and a width equal to 1.0 nm. (*a*) Use Equation 35-29 to calculate the order of magnitude of the probability that the electron will tunnel through the barrier. (*b*) Repeat your calculation for a width of 0.10 nm.

Picture the Problem The probability that the electron with a given energy will tunnel through the given barrier is given by Equation 35-29.

(*a*) Equation 35-29 is:

$$T = e^{-2\alpha a}$$

where

$$\alpha = \sqrt{\frac{2m(U_0 - E)}{\hbar^2}} = \frac{\sqrt{2m(U_0 - E)}}{\hbar}$$

Multiply the numerator and denominator of α by c to obtain:

$$\alpha = \frac{\sqrt{2mc^2(U_0 - E)}}{\hbar c}$$

where

$$\hbar c = 1.974 \times 10^{-13}\ \text{MeV} \cdot \text{m}$$

Using $m_e c^2 = 511\,\text{keV}$, evaluate T:

$$T = \exp\left\{-2(1.0 \times 10^{-9}\ \text{m})\frac{\sqrt{2(511\,\text{keV})(25\,\text{eV} - 10\,\text{eV})}}{1.974 \times 10^{-13}\ \text{MeV} \cdot \text{m}}\right\} = 5.9 \times 10^{-18} \approx \boxed{10^{-17}}$$

(*b*) Repeat with $a = 0.10$ nm:

$$T = \exp\left\{-2(0.10 \times 10^{-9}\ \text{m})\frac{\sqrt{2(511\,\text{keV})(25\,\text{eV} - 10\,\text{eV})}}{1.974 \times 10^{-13}\ \text{MeV} \cdot \text{m}}\right\} = 1.9 \times 10^{-2} \approx \boxed{10^{-2}}$$

General Problems

29 •• Eight identical non-interacting fermions are confined to an infinite two-dimensional square box of side length L. Determine the energies of the three lowest-energy states. (See Problem 22.)

Picture the Problem We can determine the energies of the state by identifying the four lowest quantum states that are occupied in the ground state and computing their combined energies. We can then find the energy difference between the ground state and the first excited state and use this information to find the energy of the excited state.

Each n, m state can accommodate only 2 particles. Therefore, in the ground state of the system of 8 fermions, the four lowest quantum states are occupied. These are:	(1,1), (1,2), (2,1) and (2,2) Note that the states (1,2) and (2,1) are distinctly different states because the x and y directions are distinguishable.

The energies are quantized to the values given by:	$E_{n_1,n_2} = 2\left(\dfrac{h^2}{8mL^2}\right)\left(n_1^2 + n_2^2\right)$

The energy of the ground state is the sum of the energies of the four lowest quantum states:

$$E_0 = E_{1,1} + E_{1,2} + E_{2,1} + E_{2,2}$$

$$= 2\left(\frac{h^2}{8mL^2}\right)\left(1^2 + 1^2\right) + 2\left(\frac{h^2}{8mL^2}\right)\left(1^2 + 2^2\right) + 2\left(\frac{h^2}{8mL^2}\right)\left(2^2 + 1^2\right)$$

$$+ 2\left(\frac{h^2}{8mL^2}\right)\left(2^2 + 2^2\right)$$

$$= 2\left(\frac{h^2}{8mL^2}\right)\left(2 + 5 + 5 + 8\right) = \frac{5h^2}{mL^2}$$

The next higher state is achieved by taking one fermion from the (2, 2) state and raising it to the next higher unoccupied state. That state is the (1, 3) state. The energy difference between the ground state and this state is:	$\Delta E = E_{1,3} - E_{2,2}$ $= \dfrac{h^2}{8mL^2}\left(1^2 + 3^2\right) - \dfrac{h^2}{8mL^2}\left(2^2 + 2^2\right)$ $= \dfrac{h^2}{4mL^2}$
Hence, the energies of the degenerate states (1,3) and (3,1) are:	$E_{1,3} = E_{3,1} = E_0 + \Delta E$ $= \dfrac{5h^2}{mL^2} + \dfrac{h^2}{4mL^2} = \dfrac{21h^2}{4mL^2}$
The three lowest energy levels are therefore:	$E_0 = \boxed{\dfrac{5h^2}{mL^2}}$ and two states of energy $E_1 = E_2 = \boxed{\dfrac{21h^2}{4mL^2}}$

37 ••• In this problem, you will derive the ground-state energy of the harmonic oscillator using the precise form of the uncertainty principle, $\Delta x \Delta p_x \geq h/2$, where Δx and Δp_x are defined to be the standard deviations $\left(\Delta x\right)^2 = \left\langle \left(x - \langle x \rangle\right)^2 \right\rangle$ and $\left(\Delta p_x\right)^2 = \left\langle \left(p_x - \langle p_x \rangle\right)^2 \right\rangle$. Proceed as follows:

1. Write the total classical energy in terms of the position x and momentum p_x using $U(x) = \frac{1}{2}m\omega_0^2 x^2$ and $K = \frac{1}{2}p_x^2/m$.

2. Show that $(\Delta x)^2 = \left\langle (x - \langle x \rangle)^2 \right\rangle = \langle x^2 \rangle - \langle x \rangle^2$ and

$(\Delta p_x)^2 = \left\langle (p_x - \langle p_x \rangle)^2 \right\rangle = \langle p_x^2 \rangle - \langle p_x \rangle^2$. (Hint: See Equations 17-34a and 17-34b.)

3. Use the symmetry of the potential energy function to argue that $\langle x \rangle$ and $\langle p_x \rangle$ must be zero, so that $(\Delta x)^2 = \langle x^2 \rangle$ and $(\Delta p_x)^2 = \langle p_x^2 \rangle$.

4. Assume that $\Delta x \Delta p_x = h/2$ to eliminate $\langle p_x^2 \rangle$ from the average energy

$\langle E \rangle = \left\langle \frac{1}{2}p_x^2/m + \frac{1}{2}m\omega_0^2 x^2 \right\rangle = \frac{1}{2}\langle p_x^2 \rangle/m + \frac{1}{2}m\omega_0^2 \langle x^2 \rangle$ and write $\langle E \rangle$ as

$\langle E \rangle = h^2/(8mZ) + \frac{1}{2}m\omega_0^2 Z$, where $Z = \langle x^2 \rangle$.

5. Set $d\langle E \rangle/dZ = 0$ to find the value of Z for which $\langle E \rangle$ is a minimum.

6. Show that the minimum energy is given by $\langle E \rangle_{min} = +\frac{1}{2}h\omega_0$.

Picture the Problem We can follow the step-by-step procedure outlined in the problem statement to show that $\langle E \rangle_{min} = +\frac{1}{2}h\omega_0$.

1. The total classical energy is:

$$E = K + U(x) = \frac{p_x^2}{2m} + \frac{1}{2}m\omega_0^2 x^2$$

Hence the average classical energy is given by:

$$\langle E \rangle = \left\langle \frac{p_x^2}{2m} + \frac{1}{2}m\omega_0^2 x^2 \right\rangle \quad (1)$$

2. Express the standard deviation of Δp_x:

$$(\Delta p_x)^2 = \left\langle (p_x - \langle p_x \rangle)^2 \right\rangle$$

$$= \left\langle p_x^2 - 2p_x\langle p_x \rangle + \langle p_x \rangle^2 \right\rangle$$

$$= \langle p_x^2 \rangle - 2\langle p_x \rangle\langle p_x \rangle + \langle p_x \rangle^2$$

$$= \langle p_x^2 \rangle - \langle p_x \rangle^2$$

Proceeding similarly for the standard deviation of Δx gives:

$$(\Delta x)^2 = \langle x^2 \rangle - \langle x \rangle^2$$

3. From the symmetry of the potential energy function we can conclude that $\langle x \rangle$ and $\langle p_x \rangle$ must be zero. Hence:

$$(\Delta x)^2 = \langle x^2 \rangle$$

and

$$(\Delta p_x)^2 = \langle p_x^2 \rangle$$

4. Rewrite equation (1) in terms of $(\Delta p_x)^2$ to obtain:

$$\langle E \rangle = \frac{\langle p_x^2 \rangle}{2m} + \tfrac{1}{2}m\omega_0^2\langle x^2 \rangle$$

$$= \frac{(\Delta p_x)^2}{2m} + \tfrac{1}{2}m\omega_0^2\langle x^2 \rangle$$

Using the uncertainty principle $(\Delta x\,\Delta p_x = h/2)$ to eliminate $\langle p_x^2 \rangle$ gives:

$$\langle E \rangle = \frac{\left(\dfrac{h}{2\Delta x}\right)^2}{2m} + \tfrac{1}{2}m\omega_0^2\langle x^2 \rangle$$

$$= \frac{h^2}{8m(\Delta x)^2} + \tfrac{1}{2}m\omega_0^2\langle x^2 \rangle$$

$$- \frac{h^2}{8m\langle x^2 \rangle} + \tfrac{1}{2}m\omega_0^2\langle x^2 \rangle$$

Letting $Z = \langle x^2 \rangle$ yields:

$$\langle E \rangle = \frac{h^2}{8mZ} + \tfrac{1}{2}m\omega_0^2 Z$$

5. Differentiate $\langle E \rangle$ with respect to Z and set this derivative equal to zero:

$$\frac{d\langle E \rangle}{dZ} = \frac{d}{dZ}\left[\frac{h^2}{8mZ} + \tfrac{1}{2}m\omega_0^2 Z \right]$$

$$= -\frac{h^2}{8mZ^2} + \tfrac{1}{2}m\omega_0^2$$

$$= 0 \text{ for extrema}$$

Solve for Z to find the value of Z that minimizes $\langle E \rangle$ (see the remark below):

$$Z = \frac{h}{2m\omega_0}$$

6. Evaluating $\langle E \rangle$ when $Z = \dfrac{h}{2m\omega_0}$ gives:

$$\langle E \rangle_{min} = \frac{h^2}{8m}\left(\frac{2m\omega_0}{h}\right) + \tfrac{1}{2}m\omega_0^2\left(\frac{h}{2m\omega_0}\right)$$

$$= \boxed{+\tfrac{1}{2}h\omega_0}$$

Remarks: All we've shown is that $Z = h/(2m\omega_0)$ is an extreme value, i.e., either a *maximum* or a *minimum*. To show that $Z = \hbar/(2m\omega_0)$ minimizes $<E>$ we must either 1) show that the second derivative of $<E>$ with respect to Z evaluated at $Z = h/(2m\omega_0)$ is positive, or 2) confirm that the graph of $<E>$ as a function of Z opens upward at $Z = h/(2m\omega_0)$.

Chapter 36
Atoms

Conceptual Problems

1 • For the hydrogen atom, as n increases, does the spacing of adjacent energy levels increase or decrease?

Determine the Concept Examination of Figure 35-4 indicates that as n increases, the spacing of adjacent energy levels decreases.

7 • For the principal quantum number $n = 4$, how many different values can the orbital quantum number ℓ take on? (*a*) 4, (*b*) 3, (*c*) 7, (*d*) 16, (*e*) 25

Determine the Concept We can find the possible values of ℓ by using the constraints on the quantum numbers n and ℓ.

The allowed values for the orbital quantum number ℓ for $n = 1, 2, 3$, and 4 are summarized in table shown to the right:

n	ℓ
1	0
2	0, 1
3	0, 1, 2
4	0, 1, 2, 3

From the table it is clear that ℓ can have 4 values. $\boxed{(a)}$ is correct.

9 •• Why is the energy of the 3s state considerably lower than the energy of the 3p state for sodium, whereas in hydrogen 3s and 3p states have essentially the same energy?

Determine the Concept The energy of a bound isolated system that consists of two oppositely charged particles, such as an electron and a proton, depends only upon the principle quantum number n. For sodium, which consists of 12 charged particles, the energy of an $n = 3$ electron depends upon the degree to which the wave function of the electron penetrates the $n = 1$ and $n = 2$ electron shells. An electron in a 3s ($n = 3$, $\ell = 0$) state penetrates these shells to a greater degree than does an electron in a 3p ($n = 3$, $\ell = 1$) state, so a 3s electron has less energy (is more tightly bound) than is a 3p electron. In hydrogen, however, the wave function of an electron in the $n = 3$ shell cannot penetrate any other electron shells because no other electron shells exist. Thus, an electron in the 3s state in hydrogen has the same energy as an electron in the 3p state in hydrogen.

19 •• The Ritz combination principle states that for any atom, one can find different spectral lines λ_1, λ_2, λ_3, and λ_4, so that $1/\lambda_1 + 1/\lambda_2 = 1/\lambda_3 + 1/\lambda_4$. Show why this is true using an energy-level diagram.

Determine the Concept The Ritz combination principle is due to the quantization of energy levels in the atom. We can use the relationship between the wavelength of the emitted photon and the difference in energy levels within the atom that results in the emission of the photon to express each of the wavelengths and then the sum of the reciprocals of the first and second wavelengths and the sum of the reciprocals of the third and fourth wavelengths.

Express the wavelengths of the spectral lines λ_1, λ_2, λ_3, and λ_4 in terms of the corresponding energy transitions:

$$\lambda_1 = \frac{hc}{E_3 - E_2}, \ \lambda_2 = \frac{hc}{E_2 - E_0}$$

$$\lambda_3 = \frac{hc}{E_3 - E_1} \ \text{and} \ \lambda_4 = \frac{hc}{E_1 - E_0}$$

Add the reciprocals of λ_1 and λ_2 to obtain:

$$\frac{1}{\lambda_1} + \frac{1}{\lambda_2} = \frac{E_3 - E_2}{hc} + \frac{E_2 - E_0}{hc}$$

$$= \frac{E_3 - E_0}{hc} \qquad (1)$$

Add the reciprocals of λ_3 and λ_4 to obtain:

$$\frac{1}{\lambda_3} + \frac{1}{\lambda_4} = \frac{E_3 - E_1}{hc} + \frac{E_1 - E_0}{hc}$$

$$= \frac{E_3 - E_0}{hc} \qquad (2)$$

Because the right-hand sides of equations (1) and (2) are equal:

$$\boxed{\frac{1}{\lambda_1} + \frac{1}{\lambda_2} = \frac{1}{\lambda_3} + \frac{1}{\lambda_4}}$$

One possible set of energy levels is shown to the right:

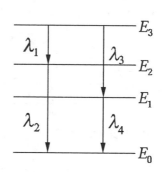

Estimation and Approximation

21 •• In laser cooling and trapping, a beam of atoms traveling in one direction are slowed by interaction with an intense laser beam in the opposite direction. The photons scatter off the atoms by resonance absorption, a process by which the incident photon is absorbed by the atom, and a short time later a photon of equal energy is emitted in a random direction. The net result of a single such scattering event is a transfer of momentum to the atom in a direction opposite to the motion of the atom, followed by a second transfer of momentum to the atom in a random direction. Thus, during photon absorption the atom loses speed, but during photon emission the change in speed of the atom is, on average, zero (because the directions of the emitted photons are random). An analogy often made to this process is that of slowing down a bowling ball by bouncing ping-pong balls off of it. (*a*) Given that the typical photon energy used in these experiments is about 1 eV, and that the typical kinetic energy of an atom in the beam is the typical kinetic energy of the atoms in a gas that has a temperature of about 500 K (a typical temperature for an oven that produces an atomic beam), estimate the number of photon-atom collisions that are required to bring an atom to rest. (The average kinetic energy of an atom is equal to $\frac{3}{2}kT$, where k is the Boltzmann constant and T is the temperature. Use this to estimate the speed of the atoms.) (*b*) Compare the Part (*a*) result with the number of ping-pong ball–bowling ball collisions that are required to bring the bowling ball to rest. Assume the typical speed of the incident ping-pong balls are all equal to the initial speed of the bowling ball.) (*c*) ^{85}Rb is a type of atom often used in cooling experiments. The wavelength of the light resonant with the cooling transition of these atoms is $\lambda = 780.24$ nm. Estimate the number of photons needed to slow down an ^{85}Rb atom from a typical thermal velocity of 300 m/s to a stop.

Picture the Problem The number of photons need to stop a ^{85}Rb atom traveling at 300 m/s is the ratio of its momentum to that of a typical photon. In Part (*b*), assume the mass of a bowling ball to be 6.0 kg and the mass of a ping pong ball to be 4.0 g.

(*a*) The number N of photon-atom collisions needed to bring an atom to rest is the ratio of the change in the momentum of the atom as it stops to the momentum brought to the collision by each photon:

$$N = \frac{\Delta p_{atom}}{p_{photon}} = \frac{mv}{\dfrac{E}{c}} = \frac{mvc}{E}$$

where m is the mass of the atom.

The kinetic energy of an atom whose temperature is T is:

$$\tfrac{1}{2}mv^2 = \tfrac{3}{2}kT \implies v = \sqrt{\frac{3kT}{m}}$$

Substituting for v gives:

$$N = \frac{mc}{E}\sqrt{\frac{3kT}{m}} = \frac{c}{E}\sqrt{3mkT}$$

For an atom whose mass is 50 u:

$$N = \frac{2.998\times10^8 \text{ m/s}}{1\text{eV}\times\dfrac{1.602\times10^{-19}\text{ J}}{\text{eV}}}\sqrt{3\left(50\,\text{u}\times\frac{1.661\times10^{-27}\text{ kg}}{\text{u}}\right)\left(1.381\times10^{-23}\text{ J/K}\right)\left(500\,\text{K}\right)}$$

$$\approx \boxed{10^5}$$

(b) The number N of ping-pong ball-bowling ball collisions needed to bring the bowling ball to rest is the ratio of the change in the momentum of the bowling ball as it stops to the momentum brought to the collision by each ping-pong ball:

$$N = \frac{\Delta p_{\text{bowling ball}}}{p_{\text{ping-pong ball}}} = \frac{m_{\text{bb}} v_{\text{bb}}}{m_{\text{ppb}} v_{\text{ppb}}}$$

Provided the speeds of the approaching bowling ball and ping-pong ball are approximately the same:

$$N = \frac{\Delta p_{\text{bowling ball}}}{p_{\text{ping-pong ball}}} \approx \frac{m_{\text{bb}}}{m_{\text{ppb}}} \approx \frac{6.0\,\text{kg}}{4.0\,\text{g}} \approx \boxed{10^3}$$

(c) The number of photons N needed to stop a ^{85}Rb atom is the ratio of the change in the momentum of the atom to the momentum brought to the collision by each photon:

$$N = \frac{\Delta p_{\text{atom}}}{p_{\text{photon}}} = \frac{mv}{\dfrac{h}{\lambda}} = \frac{mv\lambda}{h}$$

Substitute numerical values and evaluate N:

$$N = \frac{85\left(1.661 \times 10^{-27}\,\text{kg}\right)\left(300\,\text{m/s}\right)\left(780.24\,\text{nm}\right)}{6.626 \times 10^{-34}\,\text{J} \cdot \text{s}} = \boxed{5.08 \times 10^4}$$

The Bohr Model of the Hydrogen Atom

27 ••• In the center-of-mass reference frame of a hydrogen atom, the electron and nucleus have momenta that have equal magnitudes p and opposite directions. (a) Using the Bohr model, show that the total kinetic energy of the electron and nucleus can be written $K = p^2/(2\mu)$ where $\mu = m_e M/(M + m_e)$ is called the reduced mass, m_e is the mass of the electron, and M is the mass of the nucleus. (b) For the equations for the Bohr model of the atom, the motion of the nucleus can be taken into account by replacing the mass of the electron with the reduced mass. Use Equation 36-14 to calculate the Rydberg constant for a hydrogen atom that has a nucleus of mass $M = m_p$. Find the approximate value of the Rydberg constant by letting M go to infinity in the reduced mass formula. To how many figures does this approximate value agree with the actual value?
(c) Find the percentage correction for the ground-state energy of the hydrogen atom by using the reduced mass in Equation 36-16. *Remark: In general, the reduced mass for a two-body problem with masses m_1 and m_2 is given by*

$$\mu = \frac{m_1 m_2}{m_1 + m_2}.$$

Picture the Problem We can express the total kinetic energy of the electron-nucleus system as the sum of the kinetic energies of the electron and the nucleus. Rewriting these kinetic energies in terms of the momenta of the electron and nucleus will lead to $K = p^2/2m_r$.

(*a*) Express the total kinetic energy of the electron-nucleus system:

$$K = K_e + K_n$$

Express the kinetic energies of the electron and the nucleus in terms of their momenta:

$$K_e = \frac{p^2}{2m_e} \text{ and } K_n = \frac{p^2}{2M}$$

Substitute and simplify to obtain:

$$K = \frac{p^2}{2m_e} + \frac{p^2}{2M} = \frac{p^2}{2}\left(\frac{1}{m_e} + \frac{1}{M}\right)$$

$$= \frac{p^2}{2}\left(\frac{M + m_e}{m_e M}\right) = \frac{p^2}{2\left(\dfrac{m_e M}{M + m_e}\right)}$$

$$= \boxed{\frac{p^2}{2\mu}}$$

provided we let $\mu = m_e M/(M + m_e)$.

(*b*) From Equation 36-14 we have:

$$R = \frac{m_r k^2 e^4}{4\pi c\hbar^3} = C\left(\frac{m_e}{1 + \dfrac{m_e}{M}}\right) \qquad (1)$$

where

$$C = \frac{k^2 e^4}{4\pi c\hbar^3}$$

Use the Table of Physical Constants at the end of the text to obtain:

$$C = 1.204663 \times 10^{37} \text{ m}^{-1}/\text{kg}$$

For H:

$$R_H = C\left(\frac{m_e}{1 + \dfrac{m_e}{m_p}}\right)$$

Substitute numerical values and evaluate R_H :

$$R_H = \left(1.204663 \times 10^{37} \text{ m}^{-1}/\text{kg}\right) \left(\frac{9.109 \times 10^{-31} \text{ kg}}{1 + \dfrac{9.109 \times 10^{-31} \text{ kg}}{1.673 \times 10^{-27} \text{ kg}}} \right) = \boxed{1.096850 \times 10^7 \text{ m}^{-1}}$$

Let $M \to \infty$ in equation (1) to obtain $R_{H,\text{approx}}$:
$$R_{H,\text{approx}} = C m_e$$

Substitute numerical values and evaluate $R_{H,\text{approx}}$:

$$R_{H,\text{approx}} = \left(1.204663 \times 10^{37} \text{ m}^{-1}/\text{kg}\right)\left(9.109 \times 10^{-31} \text{ kg}\right) = \boxed{1.097448 \times 10^7 \text{ m}^{-1}}$$

R_H and $R_{H,\text{approx}}$ agree to three significant figures.

(c) Express the ratio of the kinetic energy K of the electron in its orbit about a stationary nucleus to the kinetic energy of the reduced-mass system K':

$$\frac{K}{K'} = \frac{\dfrac{p^2}{2m_e}}{\dfrac{p^2}{2\mu}} = \frac{\mu}{m_e} = \frac{1}{m_e}\left(\frac{m_p m_e}{m_p + m_e}\right)$$

$$= \frac{m_p}{m_p + m_e} = \frac{1}{1 + \dfrac{m_e}{m_p}}$$

Substitute numerical values and evaluate the ratio of the kinetic energies:

$$\frac{K}{K'} = \frac{1}{1 + \dfrac{9.109 \times 10^{-31} \text{ kg}}{1.673 \times 10^{-27} \text{ kg}}}$$

$$= 0.999455$$

or

$$K = 0.999455 K'$$

and the correction factor is the ratio of the masses or $\boxed{0.0545\%}$

Remarks: The correct energy is slightly less than that calculated neglecting the motion of the nucleus.

Quantum Numbers in Spherical Coordinates

33 ••   Find the minimum value of the angle θ between $\vec{L}$ and the $+z$ direction for an electron in an atom that has (a) $\ell = 1$, (b) $\ell = 4$, and (c) $\ell = 50$.

Picture the Problem The minimum angle between the z axis and $\vec{L}$ is the angle between the $\vec{L}$ vector for $m = \ell$ and the z axis.

Express the angle θ as a function of L_z and L:

$$\theta = \cos^{-1}\left(\frac{L_z}{L}\right)$$

Relate the z component of $\vec{L}$ to m_ℓ and ℓ:

$$L_z = m_\ell \hbar = \ell \hbar$$

Express the angular momentum L:

$$L = \sqrt{\ell(\ell+1)}\,\hbar$$

Substitute to obtain:

$$\theta = \cos^{-1}\left(\frac{\ell\hbar}{\sqrt{\ell(\ell+1)}\,\hbar}\right) = \cos^{-1}\left(\sqrt{\frac{\ell}{\ell+1}}\right)$$

(a) For $\ell = 1$:

$$\theta = \cos^{-1}\left(\sqrt{\frac{1}{1+1}}\right) = \boxed{45.0°}$$

(b) For $\ell = 4$:

$$\theta = \cos^{-1}\left(\sqrt{\frac{4}{4+1}}\right) = \boxed{26.6°}$$

(c) For $\ell = 50$:

$$\theta = \cos^{-1}\left(\sqrt{\frac{50}{50+1}}\right) = \boxed{8.05°}$$

Quantum Theory of the Hydrogen Atom

37 •   (a) If electron spin is not included, how many different wave functions are there corresponding to the first excited energy level $n = 2$ for a hydrogen atom? (b) List these functions by giving the quantum numbers for each state.

Picture the Problem We can use the constraints on n, ℓ and m to determine the number of different wave functions, excluding spin, corresponding to the first excited energy state of hydrogen.

For $n = 2$: $\ell = 0$ or 1

(*a*) For $\ell = 0$, $m_\ell = 0$ and we have: 1 state

For $\ell = 1$, $m_\ell = -1, 0, +1$ and we have: 3 states

Hence, for $n = 2$ we have: 4 states

(*b*) The four wave functions are summarized to the right.

n	ℓ	m_ℓ	(n, ℓ, m_ℓ)
2	0	0	(2,0,0)
2	1	−1	(2,1,−1)
2	1	0	(2,1,0)
2	1	1	(2,1,1)

General Problems

69 ••   We are often interested in finding the quantity ke^2/r in electron volts when r is given in nanometers. Show that $ke^2 = 1.44$ eV·nm.

Picture the Problem We can show that $ke^2 = 1.44$ eV·nm by solving the equation for the ground state energy of an atom for ke^2.

Express the ground state energy of an atom as a function of k, e, and a_0:

$$E_0 = \frac{ke^2}{2a_0} \Rightarrow ke^2 = 2E_0a_0$$

Substitute numerical values and evaluate ke^2:

$$ke^2 = 2(13.6\,\text{eV})(0.0529\,\text{nm})$$
$$= 1.44\,\text{eV}\cdot\text{nm}$$

Chapter 37
Molecules

Conceptual Problems

1 • Would you expect NaCl to be polar or nonpolar?

Determine the Concept Yes. Because the center of charge of the positive Na ion does not coincide with the center of charge for the negative Cl ion, the NaCl molecule has a permanent dipole moment. Hence, it is a polar molecule.

5 •• The elements on the far right column of the periodic table are sometimes called noble gases, both because they are gasses under a wide range of conditions, and because atoms of these elements almost never react with other atoms to form molecules or ionic compounds. However, atoms of noble gases can react if the resulting molecule is formed in an electronic excited state. An example is ArF. When it is formed in the excited state, it is written ArF* and is called an excimer (for <u>ex</u>cited di<u>mer</u>). Refer to Figure 37-13 and discuss how a diagram for the electronic, vibrational, and rotation energy levels of ArF and ArF* would look in which the ArF ground state is unstable and the ArF* excited state is stable. *Remark: Excimers are used in certain kinds of lasers.*

Determine the Concept The diagram would consist of a non-bonding ground state with no vibrational or rotational states for ArF (similar to the upper curve in Figure 37-4) but for ArF* there should be a bonding excited state with a definite minimum with respect to inter-nuclear separation and several vibrational states as in the excited state curve of Figure 37-13.

Energy Levels of Spectra of Diatomic Molecules

23 • The separation of the two oxygen atoms in a molecule of O_2 is actually slightly greater than the 0.100 nm used in **Error! Reference source not found.**. Furthermore, the characteristic energy of rotation E_{0r} for O_2 is 1.78×10^{-4} eV rather than the result obtained in that example. Use this value to calculate the separation distance of the two oxygen atoms.

Picture the Problem We can relate the characteristic rotational energy E_{0r} to the moment of inertia of the molecule and model the moment of inertia of the O_2 molecule as two point objects separated by a distance r.

The characteristic rotational energy of a molecule is given by:

$$E_{0r} = \frac{\hbar^2}{2I}$$

The moment of inertia of the molecule is given by:

$$I = 2M_O\left(\frac{r}{2}\right)^2 = \tfrac{1}{2}M_O r^2$$

Substitute for I to obtain:

$$E_{0r} = \frac{\hbar^2}{2\left(\frac{1}{2}M_O r^2\right)} = \frac{\hbar^2}{M_O r^2} = \frac{\hbar^2}{16 m_p r^2}$$

Solving for r gives:

$$r = \frac{\hbar}{4}\sqrt{\frac{1}{E_{0r} m_p}}$$

Substitute numerical values and evaluate r:

$$r = \left(\frac{1.055 \times 10^{-34}\ \text{J} \cdot \text{s}}{4}\right)\sqrt{\frac{1}{\left(1.78 \times 10^{-4}\ \text{eV}\right)\left(1.602 \times 10^{-19}\ \text{J/eV}\right)\left(1.673 \times 10^{-27}\ \text{kg}\right)}}$$

$$= \boxed{0.121\,\text{nm}}$$

27 ••   The equilibrium separation between the atoms of a LiH molecule is 0.16 nm. Determine the energy separation between the $\ell = 3$ and $\ell = 2$ rotational levels of this diatomic molecule.

Picture the Problem We can use the expression for the rotational energy levels of the diatomic molecule to express the energy separation ΔE between the $\ell = 3$ and $\ell = 2$ rotational levels and model the moment of inertia of the LiH molecule as two point objects separated by a distance r_0.

The energy separation between the $\ell = 3$ and $\ell = 2$ rotational levels of this diatomic molecule is given by:

$$\Delta E = E_{\ell=3} - E_{\ell=2}$$

Express the rotational energy levels $E_{\ell=3}$ and $E_{\ell=2}$ in terms of E_{0r}:

$$E_{\ell=3} = 3(3+1)E_{0r} = 12E_{0r}$$
and
$$E_{\ell=2} = 2(2+1)E_{0r} = 6E_{0r}$$

Substitute for $E_{\ell=3}$ and $E_{\ell=2}$ to obtain:

$$\Delta E = 12E_{0r} - 6E_{0r} = 6E_{0r}$$

The characteristic rotational energy of a molecule is given by $E_{0r} = \frac{\hbar^2}{2I}$.
Hence:

$$\Delta E = 6\left(\frac{\hbar^2}{2I}\right) = \frac{3\hbar^2}{I}$$

The moment of inertia of the molecule is given by:

$$I = \mu r_0^2$$

where μ is the reduced mass of the molecule.

Substituting for I yields:

$$\Delta E = \frac{3\hbar^2}{\mu r_0^2} = \frac{3\hbar^2}{\dfrac{m_{Li} m_H}{m_{Li} + m_H} r_0^2}$$

$$= \frac{3\hbar^2 (m_{Li} + m_H)}{m_{Li} m_H r_0^2}$$

Substitute numerical values and evaluate ΔE:

$$\Delta E = \frac{3(1.055 \times 10^{-34}\,\text{J} \cdot \text{s})^2 (6.94\,\text{u} + 1\,\text{u})}{(6.94\,\text{u})(1\,\text{u})(0.16\,\text{nm})^2 (1.602 \times 10^{-19}\,\text{J/eV})(1.661 \times 10^{-27}\,\text{kg/u})} = \boxed{5.6\,\text{meV}}$$

31 •• The equilibrium separation between the atoms of a CO molecule is 0.113 nm. For a molecule, such as CO, that has a permanent electric dipole moment, radiative transitions obeying the selection rule $\Delta\ell = \pm 1$ between two rotational energy levels of the same vibrational level are allowed. (That is, the selection rule $\Delta\nu = \pm 1$ does not hold.) (a) Find the moment of inertia of CO and calculate the characteristic rotational energy E_{0r} (in eV). (b) Make an energy-level diagram for the rotational levels from $\ell = 0$ to $\ell = 5$ for some vibrational level. Label the energies in electron volts, starting with $E = 0$ for $\ell = 0$. Indicate on your diagram the transitions that obey $\Delta\ell = -1$, and calculate the energies of the photons emitted. (c) Find the wavelength of the photons emitted during each transition in (b). In what region of the electromagnetic spectrum are these photons?

Picture the Problem We can find the reduced mass of CO and the moment of inertia of a CO molecule from their definitions. The energy level diagram for the rotational levels from $\ell = 0$ to $\ell = 5$ can be found using $\Delta E_{\ell,\ell-1} = 2\ell E_{0r}$. Finally, we can find the wavelength of the photons emitted for each transition using

$$\lambda_{\ell,\ell-1} = \frac{hc}{\Delta E_{\ell,\ell-1}} = \frac{hc}{2\ell \Delta E_{0r}}.$$

(a) Express the moment of inertia of CO:

$$I = \mu r_0^2$$

where μ is the reduced mass of the CO molecule.

The reduced mass of the CO molecule is given by:

$$\mu = \frac{m_C m_O}{m_C + m_O}$$

Substituting for μ gives:

$$I = \frac{m_C m_O r_0^2}{m_C + m_O}$$

Substitute numerical values ($r_0 = 0.113\,\text{nm}$) and evaluate I:

$$I = \left(\frac{(12\,\text{u})(16\,\text{u})}{12\,\text{u} + 16\,\text{u}}\right)(1.661 \times 10^{-27}\,\text{kg/u})(0.113\,\text{nm})^2 = 1.454 \times 10^{-46}\,\text{kg} \cdot \text{m}^2$$

$$= \boxed{1.45 \times 10^{-46}\,\text{kg} \cdot \text{m}^2}$$

The characteristic rotational energy E_{0r} is given by:

$$E_{0r} = \frac{\hbar^2}{2I}$$

Substitute numerical values and evaluate E_{0r}:

$$E_{0r} = \frac{(6.58 \times 10^{-16}\,\text{eV} \cdot \text{s})^2 (1.602 \times 10^{-19}\,\text{J/eV})}{2(1.454 \times 10^{-46}\,\text{kg} \cdot \text{m}^2)} = 0.2385\,\text{meV} = \boxed{0.239\,\text{meV}}$$

(*b*) The energy level diagram is shown to the right. Note that $\Delta E_{\ell,\ell-1}$, the energy difference between adjacent levels for $\Delta\ell = -1$, is $\Delta E_{\ell,\ell-1} = 2\ell E_{0r}$.

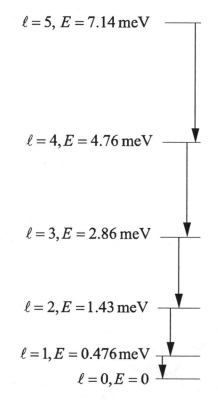

$\ell = 5, E = 7.14\,\text{meV}$

$\ell = 4, E = 4.76\,\text{meV}$

$\ell = 3, E = 2.86\,\text{meV}$

$\ell = 2, E = 1.43\,\text{meV}$

$\ell = 1, E = 0.476\,\text{meV}$

$\ell = 0, E = 0$

The energies of the photons emitted in each transition are given by $\Delta E_{\ell,\ell-1}$. Hence:

$$\Delta E_{5,4} = 7.14\,\text{meV} - 4.76\,\text{meV}$$

$$= \boxed{2.38\,\text{meV}}$$

$$\Delta E_{4,3} = 4.76\,\text{meV} - 2.86\,\text{meV}$$

$$= \boxed{1.90\,\text{meV}}$$

$$\Delta E_{3,2} = 2.86\,\text{meV} - 1.43\,\text{meV}$$

$$= \boxed{1.43\,\text{meV}}$$

$$\Delta E_{2,1} = 1.43\,\text{meV} - 0.476\,\text{meV}$$

$$= \boxed{1.25\,\text{meV}}$$

$$\Delta E_{1,0} = 0.476\,\text{meV} - 0$$

$$= \boxed{0.476\,\text{meV}}$$

(c) Express the energy difference $\Delta E_{\ell,\ell-1}$ between energy levels in terms of the frequency of the emitted radiation:

$$\Delta E_{\ell,\ell-1} = h f_{\ell,\ell-1}$$

Because $c = f_{\ell,\ell-1}\lambda_{\ell,\ell-1}$:

$$\lambda_{\ell,\ell-1} = \frac{hc}{\Delta E_{\ell,\ell-1}} = \frac{hc}{2\ell\Delta E_{0r}}$$

Substitute numerical values to obtain:

$$\lambda_{\ell,\ell-1} = \frac{\left(4.136\times10^{-15}\,\text{eV}\cdot\text{s}\right)\left(2.998\times10^{8}\,\text{m/s}\right)}{2\ell\left(0.2385\,\text{meV}\right)} = \frac{2600\,\mu\text{m}}{\ell}$$

For $\ell = 1$:

$$\lambda_{1,0} = \frac{2600\,\mu\text{m}}{1} = \boxed{2600\,\mu\text{m}}$$

For $\ell = 2$:

$$\lambda_{2,1} = \frac{2600\,\mu\text{m}}{2} = \boxed{1300\,\mu\text{m}}$$

For $\ell = 3$:

$$\lambda_{3,2} = \frac{2600\,\mu\text{m}}{3} = \boxed{867\,\mu\text{m}}$$

For $\ell = 4$:

$$\lambda_{4,3} = \frac{2600\,\mu\text{m}}{4} = \boxed{650\,\mu\text{m}}$$

For $\ell = 5$:

$$\lambda_{5,4} = \frac{2600\,\mu\text{m}}{5} = \boxed{520\,\mu\text{m}}$$

These wavelengths fall in the microwave region of the electromagnetic spectrum.

Chapter 38
Solids

Conceptual Problems

5 •   When a sample of pure copper is cooled from 300 K to 4 K, its resistivity decreases more than the resistivity of a sample of brass when it is cooled through the same temperature difference. Why?

Determine the Concept The resistivity of brass at 4 K is almost entirely due to the residual resistance (the resistance due to impurities and other imperfections of the crystal lattice). In brass, the zinc ions act as impurities in copper. In pure copper, the resistivity at 4 K is due to its residual resistance. The residual resistance is very low if the copper is very pure.

7 •   How does the change in the resistivity of a sample of copper compare with the resistivity of a sample of silicon when the temperatures of both samples increase?

Determine the Concept The resistivity of copper increases with increasing temperature; the resistivity of (pure) silicon decreases with increasing temperature because the number density of charge carriers increases.

Free Electrons in a Solid

31 ••   (*a*) Assuming that each gold atom in a sample of gold metal contributes one free electron, calculate the free-electron density in gold knowing that its atomic mass is 196.97 g/mol and its density is 19.3×10^3 kg/m^3. (*b*) If the Fermi speed for gold is 1.39×10^6 m/s, what is the Fermi energy in electron volts? (*c*) By what factor is the Fermi energy higher than the kT energy at room temperature? (*d*) Explain the difference between the Fermi energy and the kT energy.

Picture the Problem We can use $n_e = \rho V = \dfrac{\rho N_A N_{atom}}{m}$, where N_{atom} is the number of electrons per atom, to calculate the electron density of gold. The Fermi energy is given by $E_F = \frac{1}{2} m_e v_F^2$.

(*a*) The electron density of gold is given by:

$$n_e = \rho V = \frac{\rho N_A N_{atom}}{m}$$

Substitute numerical values and evaluate n_e:

$$n_e = \frac{\left(19.3\times10^3\,\frac{\text{kg}}{\text{m}^3}\right)\left(6.022\times10^{23}\,\text{atoms}\right)\left(\frac{1\,\text{e}}{1\,\text{atom}}\right)}{0.197\,\text{kg}} = \boxed{5.90\times10^{28}\,\text{e/m}^3}$$

(b) The Fermi energy is given by: $E_F = \frac{1}{2}m_e v_F^2$

Substitute numerical values and evaluate E_F:

$$E_F = \frac{1}{2}\left(9.109\times10^{-31}\,\text{kg}\right)\left(1.39\times10^6\,\text{m/s}\right)^2\left(\frac{1\,\text{eV}}{1.602\times10^{-19}\,\text{J}}\right) = \boxed{5.50\,\text{eV}}$$

(c) The factor by which the Fermi energy is higher than the kT energy at room temperature is:

$$f = \frac{E_F}{kT}$$

At room temperature $kT = 0.026$ eV. Substitute numerical values and evaluate f:

$$f = \frac{5.50\,\text{eV}}{0.026\,\text{eV}} = \boxed{212}$$

(d) The ratio E_F/kT is equal to 212 at $T = 300$ K. The Fermi energy is the energy of the most energetic conduction electron when the crystal is at absolute zero. Because no two conduction electrons can occupy the same state, the Fermi energy is quite high compared with kT. The kT energy is the energy the average conduction electron would have when the crystal is at temperature T if the electrons did not obey the exclusion principle.

33 ••   The pressure of a monatomic ideal gas is related to the average kinetic energy of the gas particles by $PV = \frac{2}{3}NE_{av}$, where N is the number of particles and E_{av} is the average kinetic energy. Use this information to calculate the pressure of the free electrons in a sample of copper in newtons per square meter, and compare your result with atmospheric pressure, which is about 10^5 N/m^2. (Note: The units are most easily handled by using the conversion factors 1 N/m^2 = 1 J/m^3 and 1 eV = 1.602 $\times$ 10^{-19} J.)

Picture the Problem We can solve $PV = \frac{2}{3}NE_{av}$ for P and substitute for E_{av} in order to express P in terms of N/V and E_F.

Solve $PV = \frac{2}{3}NE_{av}$ for P: $P = \frac{2}{3}\left(\frac{N}{V}\right)E_{av}$

Because $E_{av} = \frac{3}{5} E_F$:

$$P = \frac{2}{3}\left(\frac{N}{V}\right)\left(\frac{3}{5}E_F\right) = \frac{2}{5}\left(\frac{N}{V}\right)E_F$$

Substitute numerical values (see Table 38-1) and evaluate P:

$$P = \frac{2}{5}\left(8.47\times10^{22}\text{ electrons/cm}^3\right)\left(7.04\,\text{eV}\right)\left(1.602\times10^{-19}\text{ J/eV}\right)$$

$$= \boxed{3.82\times10^{10}\text{ N/m}^2} = 3.82\times10^{10}\text{ N/m}^2 \times \frac{1\,\text{atm}}{101.325\times10^3\text{ N/m}^2}$$

$$= \boxed{3.77\times10^5\text{ atm}}$$

Heat Capacity Due to Electrons in a Metal

35 ••   Gold has a Fermi energy of 5.53 eV. Determine the molar specific heat at constant volume and at room temperature for gold.

Picture the Problem We can use Equation 38-29 to find the molar specific heat of gold at constant volume and room temperature.

The molar specific heat is given by Equation 38-29:

$$c'_V = \frac{\pi^2 RT}{2T_F}$$

The Fermi energy is given by:

$$E_F = kT_F \Rightarrow T_F = \frac{E_F}{k}$$

Substitute for T_F to obtain:

$$c'_V = \frac{\pi^2 RkT}{2E_F}$$

Substitute numerical values and evaluate c'_V :

$$c'_V = \frac{\pi^2\left(8.314\,\text{J/mol K}\right)\left(1.381\times10^{-23}\text{ J/mol}\right)\left(\dfrac{1\,\text{eV}}{1.602\times10^{-19}\text{ J}}\right)\left(300\text{ K}\right)}{2\left(5.53\text{ eV}\right)}$$

$$= \boxed{0.192\,\text{J/(mol}\cdot\text{K)}}$$

Remarks: The value 0.192 J/mol K is for a mole of gold atoms. Since each gold atom contributes one electron to the metal, a mole of gold corresponds to a mole of electrons.

Quantum Theory of Electrical Conduction

37 ••   The resistivity of pure copper increases by approximately 1.0×10^{-8} $\Omega \cdot$m with the addition of 1 percent (by number of atoms) of an impurity distributed throughout the metal. The mean free path λ depends on both the impurity and the oscillations of the lattice ions according to the equation $1/\lambda = 1/\lambda_t + 1/\lambda_I$, where λ_t is the mean free path associated with the thermal vibrations of the ions and λ_I is the mean free path associated with the impurities. (*a*) Estimate λ_i using Equation 38-14 and the data given in Table 38-1. (*b*) If *r* is the effective radius of an impurity lattice ion seen by an electron, the scattering cross section is πr^2. Estimate this area, using the fact that *r* is related to λ_i by Equation 38-16.

Picture the Problem We can solve the resistivity equation for the mean free path and then substitute the Fermi speed for the average speed to express the mean free path as a function of the Fermi energy.

(*a*) In terms of the mean free path and the mean speed, the resistivity is:

$$\rho_i = \frac{m_e v_{av}}{ne^2 \lambda_i} = \frac{m_e u_F}{ne^2 \lambda_i} \Rightarrow \lambda_i = \frac{m_e u_F}{ne^2 \rho_i}$$

Express the Fermi speed u_F in terms of the Fermi energy E_F:

$$u_F = \sqrt{\frac{2E_F}{m_e}}$$

Substitute for u_F and simplify to obtain:

$$\lambda_i = \frac{\sqrt{2m_e E_F}}{ne^2 \rho_i}$$

Substitute numerical values (see Table 38-1) and evaluate λ_i:

$$\lambda_i = \frac{\sqrt{2(9.109 \times 10^{-31}\ \text{kg})(7.04\ \text{eV})(1.602 \times 10^{-19}\ \text{J/eV})}}{(8.47 \times 10^{28}\ \text{electrons/m}^3)(1.602 \times 10^{-19}\ \text{C})^2(1.0 \times 10^{-8}\ \Omega \cdot \text{m})} = 66.1\,\text{nm}$$

$$= \boxed{66\,\text{nm}}$$

(*b*) From Equation 38-16 we have:

$$\lambda = \frac{1}{n\pi r^2} \Rightarrow \pi r^2 = \frac{1}{n\lambda}$$

Substitute numerical values and evaluate πr^2:

$$\pi r^2 = \frac{1}{(8.47 \times 10^{28}\ \text{m}^{-3})(66.1\,\text{nm})}$$

$$= \boxed{1.8 \times 10^{-4}\,\text{nm}^2}$$

Band Theory of Solids

39 •   An electron occupies the highest energy level of the valence band in a silicon sample. What is the maximum photon wavelength that will excite the electron across the energy gap if the gap is 1.14 eV?

Picture the Problem We can relate the maximum photon wavelength to the energy gag using $\Delta E = hf = hc/\lambda$.

Express the energy gap as a function of the wavelength of the photon:

$$\Delta E = hf = \frac{hc}{\lambda} \Rightarrow \lambda = \frac{hc}{\Delta E}$$

Substitute numerical values and evaluate λ:

$$\lambda = \frac{1240\,\text{eV}\cdot\text{nm}}{1.14\,\text{eV}} = \boxed{1.09\,\mu\text{m}}$$

Semiconductor Junctions and Devices

51 ••   In Figure 38-28 for the *pnp*-transistor amplifier, suppose $R_b = 2.00$ kΩ and $R_L = 10.0$ kΩ. Suppose further that a 10.0-μA ac base current i_b generates a 0.500-mA ac collector current i_c. What is the voltage gain of the amplifier?

Picture the Problem We can use its definition to compute the voltage gain of the amplifier.

The voltage gain of the amplifier is given by:

$$\text{Voltage gain} = \frac{i_c R_L}{i_b R_b}$$

Substitute numerical values and evaluate the voltage gain:

$$\text{Voltage gain} = \frac{(0.500\,\text{mA})(10.0\,\text{k}\Omega)}{(10.0\,\mu\text{A})(2.00\,\text{k}\Omega)}$$

$$= \boxed{250}$$

The BCS Theory

57 •   (*a*) Use Equation 38-37 to calculate the superconducting energy gap for lead, and compare your result with the measured value of 2.73×10^{-3} eV. (*b*) Use the measured value to calculate the minimum value of the wavelength of a photon that has sufficient energy to break up Cooper pairs in tin ($T_c = 7.19$ K) at $T = 0$.

Picture the Problem We can calculate E_g using $E_g = 3.5kT_c$ and find the minimum value of the wavelength of a photon that has sufficient energy to break up Cooper pairs in tin at $T = 0$ using $\lambda = hc/E_g$.

(a) From Equation 38-37 we have:

$$E_g = 3.5kT_c$$

Substitute numerical values and evaluate E_g:

$$E_g = 3.5(8.62 \times 10^{-5} \text{ eV/K})(7.19 \text{ K})$$

$$= \boxed{2.17 \text{ meV}}$$

Express the ratio of E_g to $E_{g,measured}$:

$$\frac{E_g}{E_{g,measured}} = \frac{2.17 \text{ meV}}{2.73 \times 10^{-3} \text{ eV}} = 0.795$$

or

$$E_g \approx \boxed{0.8 E_{g,measured}}$$

(b) The minimum value of the wavelength of a photon that has sufficient energy to break up Cooper pairs in tin at $T = 0$ is given by:

$$\lambda = \frac{hc}{E_g}$$

Substitute numerical values and evaluate λ:

$$\lambda = \frac{1240 \text{ eV} \cdot \text{nm}}{2.73 \times 10^{-3} \text{ eV}} = \boxed{0.454 \text{ mm}}$$

The Fermi-Dirac Distribution

61 •• (a) Using the equation $E_F = \left[h^2 / (8m_e) \right] \left[3N / (\pi V) \right]^{2/3}$ (Equation 38-22a), calculate the Fermi energy for silver. (b) Determine the average kinetic energy of a free electron and (c) find the Fermi speed for silver.

Picture the Problem Equation 38-22a expresses the dependence of the Fermi energy E_F on the number density of free electrons. Once we've determined the Fermi energy for silver, we can find the average kinetic energy of a free electron from the Fermi energy for silver and then use the average kinetic energy of a free electron to find the Fermi speed for silver.

(a) From Equation 38-22a we have:

$$E_F = \frac{h^2}{8m_e} \left(\frac{3N}{\pi V} \right)^{3/2}$$

Use Table 27-1 to find the free-electron number density N/V for silver:

$$\frac{N}{V} = 5.86 \times 10^{22} \frac{\text{electrons}}{\text{cm}^3}$$

$$= 5.86 \times 10^{28} \frac{\text{electrons}}{\text{m}^3}$$

Substitute numerical values and evaluate E_F:

$$E_F = \frac{\left(6.626 \times 10^{-34}\,\text{J}\cdot\text{s}\right)^2}{8\left(9.109 \times 10^{-31}\,\text{kg}\right)}\left[\frac{3\left(5.86 \times 10^{28}\,\text{electrons/m}^3\right)}{\pi}\right]^{2/3}\left(\frac{1\,\text{eV}}{1.602 \times 10^{-19}\,\text{J}}\right)$$

$$= \boxed{5.51\,\text{eV}}$$

(b) The average kinetic energy of a free electron is given by:

$$E_{av} = \frac{3}{5}E_F$$

Substitute numerical values and evaluate E_{av}:

$$E_{av} = \frac{3}{5}(5.51\,\text{eV}) = \boxed{3.31\,\text{eV}}$$

(c) Express the Fermi energy in terms of the Fermi speed of the electrons:

$$E_F = \tfrac{1}{2}m_e v_F^2 \Rightarrow v_F = \sqrt{\frac{2E_F}{m_e}}$$

Substitute numerical values and evaluate v_F:

$$v_F = \sqrt{\frac{2(3.31\,\text{eV})}{9.109 \times 10^{-31}\,\text{kg}}\left(\frac{1.602 \times 10^{-19}\,\text{J}}{1\,\text{eV}}\right)}$$

$$= \boxed{1.08 \times 10^6\,\text{m/s}}$$

63 •• What is the probability that a conduction electron in silver will have a kinetic energy of 4.90 eV at $T = 300$ K?

Picture the Problem The probability that a conduction electron will have a given kinetic energy is given by the Fermi factor.

The Fermi factor is:

$$f(E) = \frac{1}{e^{(E-E_F)/kT} + 1}$$

Because $E_F - 4.90\,\text{eV} \gg 300k$:

$$f(4.90\,\text{eV}) = \frac{1}{0+1} = \boxed{1}$$

Chapter 39
Relativity

Conceptual Problems

1 •   The approximate total energy of a particle of mass m moving at speed $u \ll c$ is $(a)\, mc^2 + \frac{1}{2}mu^2$, $(b)\, \frac{1}{2}mu^2$, $(c)\, cmu$, $(d)\, mc^2$, $(e)\, \frac{1}{2}cmu$.

Determine the Concept The total relativistic energy E of a particle is defined to be the sum of its kinetic and rest energies.

The sum of the kinetic and rest energies of a particle is given by:

$$E = K + mc^2 = \tfrac{1}{2}mu^2 + mc^2$$

and $\boxed{(a)}$ is correct.

Estimation and Approximation

7 ••   The most distant galaxies that can be seen by the Hubble telescope are moving away from us and have a redshift parameter of about $z = 5$. (The redshift parameter z is defined as $(f - f')/f'$, where f is the frequency measured in the rest frame of the emitter, and f' is the frequency measured in the rest frame of the receiver.) (a) What is the speed of these galaxies relative to us (expressed as a fraction of the speed of light)? (b) *Hubble's law* states that the recession speed is given by the expression $v = Hx$, where v is the speed of recession, x is the distance, and H, the Hubble constant, is equal to 75 km/s/Mpc, where 1 pc = 3.26 $c\cdot$y. (The abbreviation for parsec is pc.) Estimate the distance of such a galaxy from us using the information given.

Picture the Problem (a) We can use the definition of the redshift parameter and the relativistic Doppler shift equation to show that, for light that is Doppler-shifted with respect to an observer, $v = c\left(\dfrac{u^2 - 1}{u^2 + 1}\right)$, where $u = z + 1$, and to find the ratio of v to c. In Part (b) we can solve Hubble's law for x and substitute our result from Part (a) to estimate the distance to the galaxy.

(a) The red-shift parameter is defined to be:

$$z = \frac{f_0 - f'}{f'}$$

The relativistic Doppler shift for recession is given by:

$$f' = f_0 \sqrt{\frac{1 - v/c}{1 + v/c}}$$

Substitute for f' and simplify to obtain:	$$z = \frac{f_0 - f_0\sqrt{\dfrac{11v/c}{1+v/c}}}{f_0\sqrt{\dfrac{1-v/c}{1+v/c}}} = \sqrt{\frac{1+v/c}{1-v/c}} - 1$$
Letting $u = z + 1$ and simplifying yields:	$$u = z + 1 = \sqrt{\frac{1+v/c}{1-v/c}} \Rightarrow \frac{v}{c} = \frac{u^2-1}{u^2+1}$$
Substitute for u to express v/c as a function of z:	$$\frac{v}{c} = \frac{(z+1)^2-1}{(z+1)^2+1}$$
Substituting the numerical value of z and evaluating v/c gives:	$$\frac{v}{c} = \frac{(5+1)^2-1}{(5+1)^2+1} = \boxed{0.946}$$
(b) Solving Hubble's law for x yields:	$$x = \frac{v}{H}$$
Substitute numerical values and evaluate x:	$$x = \frac{0.946c}{H} = \frac{0.946c}{75\dfrac{\text{km/s}}{\text{Mpc}}} \times \frac{3.26\times10^6\, c\cdot\text{y}}{\text{Mpc}}$$ $$= \boxed{1.23\times10^{10}\, c\cdot\text{y}}$$

Time Dilation and Length Contraction

13 •• Unobtainium (Un) is an unstable particle that decays into normalium (Nr) and standardium (St) particles. (a) An accelerator produces a beam of Un that travels to a detector located 100 m away from the accelerator. The particles travel with a velocity of $v = 0.866c$. How long do the particles take (in the laboratory frame) to get to the detector? (b) By the time the particles get to the detector, half of the particles have decayed. What is the half-life of Un? (*Note:* half-life as it would be measured in a frame moving with the particles) (c) A new detector is going to be used, which is located 1000 m away from the accelerator. How fast should the particles be moving if half of the particles are to make it to the new detector?

Picture the Problem The time required for the particles to reach the detector, as measured in the laboratory frame of reference is the ratio of the distance they must travel to their speed. The half life of the particles is the trip time as measured in a frame traveling with the particles. We can find the speed at which the particles must move if they are to reach the more distant detector by equating their half life to the ratio of the distance to the detector in the particle's frame of reference to their speed.

(*a*) The time required to reach the detector is the ratio of the distance to the detector and the speed with which the particles are traveling:

$$\Delta t = \frac{\Delta x}{v} = \frac{\Delta x}{0.866c}$$

Substitute numerical values and evaluate Δt:

$$\Delta t = \frac{100\,\text{m}}{0.866\left(2.998\times10^8\,\text{m/s}\right)}$$

$$= \boxed{0.385\,\mu s}$$

(*b*) The half life is the trip time as measured in a frame traveling with the particles:

$$\Delta t' = \frac{\Delta t}{\gamma} = \Delta t\sqrt{1-\left(\frac{v}{c}\right)^2}$$

Substitute numerical values and evaluate $\Delta t'$:

$$\Delta t' = \left(0.385\,\mu s\right)\sqrt{1-\left(\frac{0.866c}{c}\right)^2}$$

$$= \boxed{0.193\,\mu s}$$

(*c*) In order for half the particles to reach the detector:

$$\Delta t' = \frac{\Delta x'}{\gamma v} = \frac{\Delta x'\sqrt{1-\left(\frac{v}{c}\right)^2}}{v}$$

where $\Delta x'$ is the distance to the new detector.

Rewrite this expression to obtain:

$$\frac{v}{\sqrt{1-\left(\frac{v}{c}\right)^2}} = \frac{\Delta x'}{\Delta t'}$$

Squaring both sides of the equation yields:

$$\frac{v^2}{1-\left(\frac{v}{c}\right)^2} = \left(\frac{\Delta x'}{\Delta t'}\right)^2$$

Substitute numerical values for $\Delta x'$ and $\Delta t'$ and simplify to obtain:

$$\frac{v^2}{1-\left(\dfrac{v}{c}\right)^2} = \left(\frac{1000\,\text{m}}{0.193\,\mu s}\right)^2 = (17.3c)^2$$

Divide both sides of the equation by c^2 to obtain:

$$\frac{\dfrac{v^2}{c^2}}{1-\left(\dfrac{v}{c}\right)^2} = (17.3)^2$$

Solving this equation for v^2/c^2 gives:

$$\frac{v^2}{c^2} = \frac{(17.3)^2}{1 + (17.3)^2} = 0.9967$$

Finally, solving for v yields:

$$v = \boxed{0.998c}$$

The Lorentz Transformation, Clock Synchronization, and Simultaneity

17 •• A spaceship of proper length $L_p = 400$ m moves past a transmitting station at a speed of $0.760c$. (The transmitting station broadcasts signals that travel at the speed of light.) A clock is attached to the nose of the spaceship and a second clock is attached to the transmitting station. The instant that the nose of the spaceship passes the transmitter, the clock attached to the transmitter and the clock attached to the nose of the spaceship are set equal to zero. The instant that the tail of the spaceship passes the transmitter a signal is sent by the transmitter that is subsequently detected by a receiver in the nose of the spaceship. (*a*) When, according to the clock attached to the nose of spaceship, is the signal sent? (*b*) When, according to the clocks attached to the nose of spaceship, is the signal received? (*c*) When, according to the clock attached to the transmitter, is the signal received by the spaceship? (*d*) According to an observer that works at the transmitting station, how far from the transmitter is the nose of the spaceship when the signal is received?

Picture the Problem Let S be the reference frame of the spaceship and S' be that of Earth (transmitter station). Let event A be the emission of the light pulse and event B the reception of the light pulse at the nose of the spaceship. In (*a*) and (*c*) we can use the classical distance, rate, and time relationship and in (*b*) and (*d*) we can apply the inverse Lorentz transformations.

(*a*) In both S and S' the pulse travels at the speed c. Thus:

$$t_A = \frac{L_p}{v} = \frac{400\,\text{m}}{0.760c} = \boxed{1.76\,\mu s}$$

(b) The inverse time transformation is:

$$t_B' = \gamma\left(t - \frac{vx}{c^2}\right)$$

where

$$\gamma = \frac{1}{\sqrt{1 - \dfrac{v^2}{c^2}}} = \frac{1}{\sqrt{1 - \dfrac{(0.760c)^2}{c^2}}} = 1.54$$

Substitute numerical values and evaluate t_B':

$$t_B' = (1.54)\left(3.09\,\mu s - \frac{(-0.760c)(400\,m)}{c^2}\right)$$

$$= (1.54)\left(3.09\,\mu s - \frac{(-0.760)(400\,m)}{2.998 \times 10^8\,m/s}\right)$$

$$= \boxed{6.32\,\mu s}$$

(c) The elapsed time, according to the clock on the ship is:

$$t_B = t_{\substack{\text{pulse to travel} \\ \text{length of ship}}} + t_A$$

Find the time of travel of the pulse to the nose of the ship:

$$t_{\substack{\text{pulse to travel} \\ \text{length of ship}}} = \frac{400\,m}{2.998 \times 10^8\,m/s}$$

$$= 1.33\,\mu s$$

Substitute numerical values and evaluate t_B:

$$t_B = 1.33\,\mu s + 1.76\,\mu s = \boxed{3.09\,\mu s}$$

(d) The inverse transformation for x is:

$$x' = \gamma(x - vt)$$

Substitute numerical values and evaluate x':

$$x' = (1.54)\left[400\,m - (-0.760)(2.998 \times 10^8\,m/s)(3.09 \times 10^{-6}\,s)\right] = \boxed{1.70\,km}$$

The Relativistic Doppler Shift

27 ••   A clock is placed in a satellite that orbits Earth with an orbital period of 90 min. By what time interval will this clock differ from an identical clock on Earth after 1.0 y? (Assume that special relativity applies and neglect general relativity.)

Picture the Problem Due to its motion, the orbiting clock will run more slowly than the Earth-bound clock. We can use Kepler's third law to find the radius of the satellite's orbit in terms of its period, the definition of speed to find the orbital speed of the satellite from the radius of its orbit, and the time dilation equation to find the difference δ in the readings of the two clocks.

Express the time δ lost by the clock:

$$\delta = \Delta t - \Delta t_p = \Delta t - \frac{\Delta t}{\gamma} = \Delta t \left(1 - \frac{1}{\gamma}\right)$$

Because $v \ll c$, we can use Part (b) of Problem 10:

$$\frac{1}{\gamma} \approx 1 - \frac{1}{2}\frac{v^2}{c^2}$$

Substitute to obtain:

$$\delta = \Delta t \left[1 - \left(1 - \frac{1}{2}\frac{v^2}{c^2}\right)\right] = \frac{1}{2}\frac{v^2}{c^2}\Delta t \quad (1)$$

Express the square of the speed of the satellite in its orbit:

$$v^2 = \left(\frac{2\pi r}{T}\right)^2 = \frac{4\pi^2 r^2}{T^2} \quad (2)$$

where T is its period and r is the radius of its (assumed) circular orbit.

Use Kepler's third law to relate the period of the satellite to the radius of its orbit about Earth:

$$T^2 = \frac{4\pi^2}{GM_E}r^3 = \frac{4\pi^2}{gR_E^2}r^3 \Rightarrow r = \sqrt[3]{\frac{gR_E^2 T^2}{4\pi^2}}$$

Substitute numerical values and evaluate r:

$$r = \sqrt[3]{\frac{(9.81\,\text{m/s}^2)(6370\,\text{km})^2(90\,\text{min} \times 60\,\text{s/min})^2}{4\pi^2}} = 6.65 \times 10^6 \text{ m}$$

Substitute numerical values in equation (2) and evaluate v^2:

$$v^2 = \frac{4\pi^2(6.65 \times 10^6 \text{ m})^2}{(90\,\text{min} \times 60\,\text{s/min})^2}$$
$$= 5.99 \times 10^7 \text{ m}^2/\text{s}^2$$

Finally, substitute for v^2 in equation (1) and evaluate δ:

$$\delta = \frac{1}{2}\frac{(5.99 \times 10^7 \text{ m}^2/\text{s}^2)(1.0\,\text{y} \times 31.56\,\text{Ms/y})}{(2.998 \times 10^8 \text{ m/s})^2} = \boxed{11\,\text{ms}}$$

31 •• A particle moves with speed $0.800c$ in the $+x''$ direction along the x'' axis of frame S'', which moves with the same speed and in the same direction along the x' axis relative to frame S'. Frame S' moves with the same speed and in the same direction along the x axis relative to frame S. (a) Find the speed of the particle relative to frame S'. (b) Find the speed of the particle relative to frame S.

Picture the Problem We can apply the inverse velocity transformation equation to express the speed of the particle relative to both frames of reference.

(a) Express u_x' in terms of u_x'':

$$u_x' = \frac{u_x'' + v}{1 + \dfrac{vu_x''}{c^2}}$$

where v of S', relative to S'', is $0.800c$.

Substitute numerical values and evaluate u_x':

$$u_x' = \frac{0.800c + 0.800c}{1 + \dfrac{(0.800c)^2}{c^2}} = \frac{1.60c}{1.64}$$

$$= \boxed{0.976c}$$

(b) Express u_x in terms of u_x':

$$u_x = \frac{u_x' + v}{1 + \dfrac{vu_x'}{c^2}} \quad \text{where } v, \text{ the speed of } S,$$

relative to S', is $0.800c$.

Substitute numerical values and evaluate u_x:

$$u_x = \frac{0.976c + 0.800c}{1 + \dfrac{(0.800c)(0.976c)}{c^2}} = \frac{1.776c}{1.781}$$

$$= \boxed{0.997c}$$

Relativistic Momentum and Relativistic Energy

37 ••   In reference frame S', two protons, each moving at $0.500c$, approach each other head-on. (a) Calculate the total kinetic energy of the two protons in frame S'. (b) Calculate the total kinetic energy of the protons as seen in reference frame S, which is moving with speed $0.500c$ relative to S' so that one of the protons is at rest.

Picture the Problem The total kinetic energy of the two protons in Part (a) is the sum of their kinetic energies and is given by $K = 2(\gamma - 1)E_0$. Part (b) differs from Part (a) in that we need to find the speed of the moving proton relative to frame S.

(*a*) The total kinetic energy of the protons in frame S' is given by:

$$K = 2(\gamma - 1)E_0$$

Substitute for γ and E_0 and evaluate K:

$$K = 2\left(\frac{1}{\sqrt{1 - \frac{(0.500c)^2}{c^2}}} - 1\right)(938.28\,\text{MeV})$$

$$= \boxed{290\,\text{MeV}}$$

(*b*) The kinetic energy of the moving proton in frame S is given by:

$$K = (\gamma - 1)E_0 \qquad (1)$$

where

$$\gamma = \frac{1}{\sqrt{1 - \frac{uv}{c^2}}}$$

Express the speed u of the proton in frame S:

$$u = \frac{u_x' + v}{1 + \frac{vu_x'}{c^2}}$$

Substitute numerical values and evaluate u:

$$u = \frac{0.500c + 0.500c}{1 + \frac{(0.500c)(0.500c)}{c^2}} = 0.800c$$

Evaluate γ:

$$\gamma = \frac{1}{\sqrt{1 - \frac{(0.800c)(0.800c)}{c^2}}} = 1.67$$

Substitute numerical values in equation (1) and evaluate K:

$$K = (1.67 - 1)(938.28\,\text{MeV})$$

$$= \boxed{629\,\text{MeV}}$$

General Problems

45 •• Frames S and S' are moving relative to each other along the x and x' axes (which superpose). Observers at rest in the two frames set their clocks to $t = 0$ when the two origins coincide. In frame S, event 1 occurs at $x_1 = 1.0\ c \cdot y$ and $t_1 = 1.00$ y and event 2 occurs at $x_2 = 2.0\ c \cdot y$ and $t_2 = 0.50$ y. These events occur simultaneously in frame S'. (*a*) Find the magnitude and direction of the velocity of S' relative to S. (*b*) At what time do both these events occur as measured in S'?

Picture the Problem We can use Equation 39-12, the inverse time transformation equation, to relate the elapsed times and separations of the events in the two systems to the velocity of S' relative to S. We can use this same relationship in Part (*b*) to find the time at which these events occur as measured in S'.

(*a*) Use Equation 39-12 to obtain:

$$\Delta t' = t_2' - t_1' = \gamma \left[(t_2 - t_1) - \frac{v}{c^2}(x_2 - x_1) \right]$$

$$= \gamma \left[\Delta t - \frac{v}{c^2} \Delta x \right]$$

Because the events occur simultaneously in frame S', $\Delta t' = 0$ and:

$$0 = \Delta t - \frac{v}{c^2} \Delta x \Rightarrow v = \frac{c^2 \Delta t}{\Delta x}$$

Substitute for Δt and Δx and evaluate v:

$$v = \frac{c^2(0.50\,\text{y} - 1.00\,\text{y})}{2.0\,c \cdot \text{y} - 1.00\,c \cdot \text{y}} = \boxed{-0.50c}$$

Because $\Delta t = t_2 - t_1 = -0.50\,\text{y}$:

$$\boxed{S' \text{ moves in the} - x \text{ direction.}}$$

(*b*) Use the inverse time transformation to obtain:

$$t_2' = \gamma\left(t_2 - \frac{v x_2}{c^2}\right) = \frac{t_2 - \frac{v x_2}{c^2}}{\sqrt{1 - \frac{v^2}{c^2}}}$$

Substitute numerical values and evaluate t_2' and t_1':

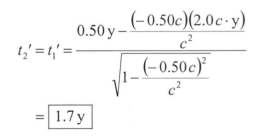

$$= \boxed{1.7\,\text{y}}$$

49 ••   Using a simple thought experiment, Einstein showed that there is mass associated with electromagnetic radiation. Consider a box of length L and mass M resting on a frictionless surface. Attached to the left wall of the box is a light source that emits a directed pulse of radiation of energy E, which is completely absorbed at the right wall of the box. According to classical electromagnetic theory, this radiation carries momentum of magnitude $p = E/c$ (Equation 32-13). The box recoils when the pulse is emitted by the light source. (*a*) Find the recoil velocity of the box so that momentum is conserved when the light is emitted. (Because p is small and M is large, you may use classical mechanics.) (*b*) When the light is absorbed at the right wall of the box the box stops, so the total momentum of the system remains zero. If we neglect the very small velocity of the box, the time it takes for the radiation to travel across the box is $\Delta t = L/c$. Find the distance moved by the box in this time. (*c*) Show that if the center of mass of the system is to remain at the same place, the radiation must carry mass $m = E/c^2$.

Picture the Problem We can use conservation of energy to express the recoil velocity of the box and the relationship between distance, speed, and time to find the distance traveled by the box in time $\Delta t = L/c$. Equating the initial and final locations of the center of mass will allow us to show that the radiation must carry mass $m = E/c^2$.

(*a*) Apply conservation of momentum to obtain:

$$\frac{E}{c} + Mv = p_i = 0 \Rightarrow v = \boxed{-\frac{E}{Mc}}$$

(*b*) The distance traveled by the box in time $\Delta t = L/c$ is:

$$d = v\Delta t = \frac{vL}{c}$$

Substitute for v from (*a*) to obtain:

$$d = \frac{L}{c}\left(-\frac{E}{Mc}\right) = \boxed{-\frac{LE}{Mc^2}}$$

(*c*) Let $x = 0$ be at the center of the box and let the mass of the photon be m. Then initially the center of mass is at:

$$x_{CM} = \frac{-\frac{1}{2}mL}{M+m}$$

When the photon is absorbed at the other end of the box, the center of mass is at:

$$x_{CM} = \frac{\left[\frac{-MEL}{Mc^2} + m\left(\frac{1}{2}L - \frac{EL}{Mc^2}\right)\right]}{M+m}$$

Because no external forces act on the system, these expressions for x_{CM} must be equal:

$$\frac{-\frac{1}{2}mL}{M+m} = \frac{\left[\frac{-MEL}{Mc^2} + m\left(\frac{1}{2}L - \frac{EL}{Mc^2}\right)\right]}{M+m}$$

Solving for m yields:

$$m = \frac{E}{c^2\left(1 - \frac{E}{Mc^2}\right)}$$

Because Mc^2 is of the order of 10^{16} J and $E = hf$ is of the order of 1 J for reasonable values of f, $E/Mc^2 \ll 1$ and:

$$m = \boxed{\frac{E}{c^2}}$$

51 ••• When a moving particle that has a kinetic energy greater than the threshold kinetic energy K_{th} strikes a stationary target particle, one or more particles may be created in the inelastic collision. Show that the threshold kinetic energy of the moving particle is given by

$$K_{th} = \frac{\left(\Sigma m_{in} + \Sigma m_{fin}\right)\left(\Sigma m_{fin} + \Sigma m_{in}\right)c^2}{2m_{target}}$$

Here Σm_{in} is the sum of the masses of the particles prior to the collision, Σm_{fin} is the sum of the masses of the particles following the collision, and m_{target} is the mass of the target particle. Use this expression to determine the threshold kinetic energy of protons incident on a stationary proton target for the production of a proton–antiproton pair; compare your result with the result of Problem 38.

Picture the Problem Let m_i denote the mass of the incident (projectile) particle. Then $\Sigma m_{in} = m_i + m_{target}$ and we can use this expression to determine the threshold kinetic energy of protons incident on a stationary proton target for the production of a proton–antiproton pair.

Consider the situation in the center of mass reference frame. At threshold we have:	$E^2 - p^2c^2 = \sum m_{fin}c^2$ Note that this is a relativistically invariant expression.
In the laboratory frame, the target is at rest so:	$E_{target} = E_t = \dot{E}_{t,0}$
We can, therefore, write:	$\left(E_i + E_{t,0}\right)^2 - p_i^2c^2 = \left(\sum m_{fin}c^2\right)^2$
For the incident particle:	$E_i^2 - p_i^2c^2 = E_{i,0}^2$ and $E_i = E_{i,0} + K_{th}$ where K_{th} is the threshold kinetic energy of the incident particle in the laboratory frame.
Express K_{th} in terms of the rest energies:	$\left(E_{t,0} + E_{i,0}\right)^2 + 2K_{th}E_{t,0} = \left(\sum m_{fin}c^2\right)^2$ where $E_{t,0} + E_{i,0} = \sum m_{fin}c^2$ and $E_{t,0} = m_{target}c^2$

Substitute to obtain:

$$\left(\sum m_{\text{fin}}c^2\right)^2 + 2K_{\text{th}}m_{\text{target}}c^2 = \left(\sum m_{\text{fin}}c^2\right)^2$$

Solving for K_{th} gives:

$$K_{\text{th}} = \boxed{\frac{\left(\sum m_{\text{in}} + \sum m_{\text{fin}}\right)\left(\sum m_{\text{fin}} - \sum m_{\text{in}}\right)c^2}{2m_{\text{target}}}}$$

For the creation of a proton - antiproton pair in a proton - proton collision:

$$\sum m_{\text{in}} = 2m_{\text{p}}, \quad \sum m_{\text{fin}} = 4m_{\text{p}} \text{ and}$$
$$m_{\text{target}} = m_{\text{p}}$$

Substituting for the sums and simplifying yields:

$$K_{\text{th}} = \frac{\left(2m_{\text{p}} + 4m_{\text{p}}\right)\left(4m_{\text{p}} - 2m_0\right)c^2}{2m_{\text{p}}}$$
$$= \frac{\left(6m_{\text{p}}\right)\left(2m_{\text{p}}\right)c^2}{2m_{\text{p}}} = \boxed{6m_{\text{p}}c^2}$$

in agreement with Problem 38.

55 ••• For the special case of a particle moving with speed u along the y axis in frame S, show that its momentum and energy in frame S' are related to its momentum and energy in S by the transformation equations

$$p_x' = \gamma\left(p_x - \frac{vE}{c^2}\right), \quad p_y' = p_y, \quad p_z' = p_z, \text{ and } \frac{E'}{c} = \gamma\left(\frac{E}{c} - \frac{vp_x}{c}\right).$$

Compare these equations with the Lorentz transformation equations for x', y', z', and t'. Notice that that the quantities p_x, p_y, p_z, and E/c transform in the same way as do x, y, z, and ct.

Picture the Problem We can use the expressions for $\vec{p}$ and E in S together with the relations we wish to verify and the inverse velocity transformation equations to establish the condition $u'^2 = \left(u_x'\right)^2 + \left(u_y'\right)^2 + \left(u_z'\right)^2 = v^2 + \frac{u^2}{\gamma^2}$ and then use this result to verify the given expressions for p_x', p_y', p_z' and E'/c.

In any inertial frame the momentum and energy are given by:

$$\vec{p} = \frac{m\vec{u}}{\sqrt{1 - \dfrac{u^2}{c^2}}} \text{ and } E = \frac{mc^2}{\sqrt{1 - \dfrac{u^2}{c^2}}}$$

where $\vec{u}$ is the velocity of the particle and u is its speed.

The components of $\vec{p}$ in S are:

$$p_x = \frac{mu_x}{\sqrt{1 - \dfrac{u^2}{c^2}}} , \quad p_y = \frac{mu_y}{\sqrt{1 - \dfrac{u^2}{c^2}}} , \text{ and}$$

$$p_z = \frac{mu_z}{\sqrt{1 - \dfrac{u^2}{c^2}}}$$

Because $u_x = u_z = 0$ and $u_y = u$:

$$p_x = p_z = 0 \text{ and } p_y = \frac{mu}{\sqrt{1 - \dfrac{u^2}{c^2}}}$$

Substituting zeros for p_x and p_z in the relations we are trying to show yields:

$$p_x' = \gamma\left(0 - \frac{vE}{c^2}\right) = -\gamma\frac{vE}{c^2} , \quad p_y' = p_y ,$$

$$p_z' = 0 , \text{ and}$$

$$\frac{E'}{c} = \gamma\left(\frac{E}{c} - 0\right) = \gamma\frac{E}{c}$$

In S' the momentum components are:

$$p_x' = \frac{mu_x'}{\sqrt{1 - \dfrac{u'^2}{c^2}}} , \quad p_y' = \frac{mu_y'}{\sqrt{1 - \dfrac{u'^2}{c^2}}} , \text{ and}$$

$$p_z' = \frac{mu_z'}{\sqrt{1 - \dfrac{u'^2}{c^2}}}$$

The inverse velocity transformations are:

$$u_x' = \frac{u_x - v}{\sqrt{1 - \dfrac{vu_x}{c^2}}} , \quad u_y' = \frac{u_y}{\sqrt{1 - \dfrac{vu_y}{c^2}}} , \text{ and}$$

$$u_z' = \frac{u_z}{\sqrt{1 - \dfrac{vu_z}{c^2}}}$$

Substitute $u_x = u_z = 0$ and $u_y = u$ to obtain:

$$u_x' = -v , \quad u_y' = \gamma u , \text{ and } u_z' = 0$$

Thus:

$$u'^2 = \left(u_x'\right)^2 + \left(u_y'\right)^2 + \left(u_z'\right)^2$$

$$= v^2 + \frac{u^2}{\gamma^2}$$

First we verify that $p_z' = p_z = 0$:

$$p_z' = \frac{m(0)}{\sqrt{1 - \dfrac{u'^2}{c^2}}} = p_z = \boxed{0}$$

Next we verify that $p_y' = p_y$:

$$p_y' = \frac{mu_y'}{\sqrt{1 - \dfrac{u'^2}{c^2}}} = \frac{mu}{\gamma \sqrt{1 - \dfrac{v^2}{c^2} - \dfrac{u^2}{\gamma^2 c^2}}} = \frac{mu}{\sqrt{1 - \dfrac{u^2}{c^2}}} \; \frac{\sqrt{1 - \dfrac{u^2}{c^2}}}{\gamma \sqrt{1 - \dfrac{v^2}{c^2} - \dfrac{u^2}{\gamma^2 c^2}}}$$

$$= \frac{mu}{\sqrt{1 - \dfrac{u^2}{c^2}}} \sqrt{\frac{\left(1 - \dfrac{u^2}{c^2}\right)\left(1 - \dfrac{v^2}{c^2}\right)}{1 - \dfrac{v^2}{c^2} - \dfrac{u^2}{c^2}\left(1 - \dfrac{v^2}{c^2}\right)}} = p_y \sqrt{\frac{1 - \dfrac{v^2}{c^2} - \dfrac{u^2}{c^2}\left(1 - \dfrac{v^2}{c^2}\right)}{1 - \dfrac{v^2}{c^2} - \dfrac{u^2}{c^2}\left(1 - \dfrac{v^2}{c^2}\right)}}$$

$$= \boxed{p_y}$$

Next, we verify that $p_x' = \gamma\left(p_x - \dfrac{vE}{c^2}\right)$:

$$p_x' = \frac{mu_x'}{\sqrt{1 - \dfrac{u'^2}{c^2}}} = \frac{-mv}{\gamma \sqrt{1 - \dfrac{v^2}{c^2} - \dfrac{u^2}{\gamma^2 c^2}}} = -\frac{\gamma v}{c^2} \; \frac{mc^2}{\sqrt{1 - \dfrac{u^2}{c^2}}} \; \frac{\gamma^{-1}\sqrt{1 - \dfrac{u^2}{c^2}}}{\sqrt{1 - \dfrac{v^2}{c^2} - \dfrac{u^2}{\gamma^2 c^2}}}$$

$$= -\frac{\gamma v}{c^2} E \sqrt{\frac{\left(1 - \dfrac{u^2}{c^2}\right)\left(1 - \dfrac{v^2}{c^2}\right)}{1 - \dfrac{v^2}{c^2} - \dfrac{u^2}{c^2}\left(1 - \dfrac{v^2}{c^2}\right)}} = -\frac{\gamma v}{c^2} E \sqrt{\frac{1 - \dfrac{v^2}{c^2} - \dfrac{u^2}{c^2}\left(1 - \dfrac{v^2}{c^2}\right)}{1 - \dfrac{v^2}{c^2} - \dfrac{u^2}{c^2}\left(1 - \dfrac{v^2}{c^2}\right)}}$$

$$= \boxed{-\frac{\gamma v}{c^2} E}$$

Finally, we verify that $\dfrac{E'}{c} = \gamma\left(\dfrac{E}{c} - \dfrac{vp_x}{c}\right) = \gamma\dfrac{E}{c}$, or $E' = \gamma E$:

$$E' = \frac{mc^2}{\sqrt{1-\dfrac{u'^2}{c^2}}} = \frac{\gamma mc^2}{\sqrt{1-\dfrac{u^2}{c^2}}} \frac{\gamma^{-1}\sqrt{1-\dfrac{u^2}{c^2}}}{\sqrt{1-\dfrac{u'^2}{c^2}}} = \gamma E \frac{\gamma^{-1}\sqrt{1-\dfrac{u^2}{c^2}}}{\sqrt{1-\dfrac{v^2}{c^2}-\dfrac{u^2}{\gamma^2 c^2}}}$$

$$= \gamma E \sqrt{\frac{\left(1-\dfrac{u^2}{c^2}\right)\left(1-\dfrac{v^2}{c^2}\right)}{1-\dfrac{v^2}{c^2}-\dfrac{u^2}{c^2}\left(1-\dfrac{v^2}{c^2}\right)}} = \gamma E \sqrt{\frac{1-\dfrac{v^2}{c^2}-\dfrac{u^2}{c^2}\left(1-\dfrac{v^2}{c^2}\right)}{1-\dfrac{v^2}{c^2}-\dfrac{u^2}{c^2}\left(1-\dfrac{v^2}{c^2}\right)}}$$

$$= \boxed{\gamma E}$$

The x, y, z, and t transformation equations are:

$x' = \gamma(x - vt)$, $y' = y$, $z' = z$

and

$t' = \gamma\left(t - \dfrac{vx}{c^2}\right)$

The x, y, z, and ct transformation equations are:

$x' = \gamma\left(x - \dfrac{v}{c}ct\right)$, $y' = y$, $z' = z$

and

$ct' = \gamma\left(ct - \dfrac{v}{c}x\right)$

The p_x, p_y, p_z, and E/c transformation equations are:

$p_x' = \gamma\left(p_x - \dfrac{v}{c}\dfrac{E}{c}\right)$, $p_y' = p_y$, $p_z' = p_z$

and

$\dfrac{E'}{c} = \gamma\left(\dfrac{E}{c} - \dfrac{v}{c}p_x\right)$

Note that the transformation equations for x, y, z, and ct and the transformation equations for p_x, p_y, p_z, and E/c are identical.

Chapter 40
Nuclear Physics

Properties of Nuclei

15 • Calculate the binding energy and the binding energy per nucleon from the masses given in Table 40-1 for (a) ^{12}C, (b) ^{56}Fe, and (c) ^{238}U.

Picture the Problem To find the binding energy of a nucleus we add the mass of its neutrons to the mass of its protons and then subtract the mass of the nucleus and multiply by c^2. To convert to MeV we multiply this result by 931.5 MeV/u. The binding energy per nucleon is the ratio of the binding energy to the mass number of the nucleus.

(a) For ^{12}C, $Z = 6$ and $N = 6$. Add the mass of the neutrons to that of the protons:

$$6m_p + 6m_n = 6 \times 1.007\,825\,u + 6 \times 1.008\,665\,u = 12.098\,940\,u$$

Subtract the mass of ^{12}C from this result:

$$\left(6m_p + 6m_n\right) - m_{^{12}C} = 12.098\,940\,u - 12\,u = 0.098\,940\,u$$

Multiply the mass difference by c^2 and convert to MeV:

$$E_b = \left(\Delta m\right)c^2 = \left(0.098\,940\,u\right)c^2\left(\frac{931.5\,\text{MeV}/c^2}{u}\right) = \boxed{92.2\,\text{MeV}}$$

and the binding energy per nucleon is $\dfrac{E_b}{A} = \dfrac{92.2\,\text{MeV}}{12} = \boxed{7.68\,\text{MeV}}$

(b) For ^{56}Fe, $Z = 26$ and $N = 30$. Add the mass of the neutrons to that of the protons:

$$26m_p + 30m_n = 26 \times 1.007\,825\,u + 30 \times 1.008\,665\,u = 56.463\,400\,u$$

Subtract the mass of ^{56}Fe from this result:

$$\left(26m_p + 30m_n\right) - m_{^{12}C} = 56.463\,400\,u - 55.934\,942\,u = 0.528\,458\,u$$

Multiply the mass difference by c^2 and convert to MeV:

$$E_b = (\Delta m)c^2 = (0.528\ 458\ \text{u})c^2 \left(\frac{931.5\ \text{MeV}/c^2}{\text{u}}\right) = \boxed{492\ \text{MeV}}$$

and the binding energy per nucleon is $\dfrac{E_b}{A} = \dfrac{492\ \text{MeV}}{56} = \boxed{8.79\ \text{MeV}}$

(c) For ^{238}U, $Z = 92$ and $N = 146$. Add the mass of the neutrons to that of the protons:

$$92 m_p + 146 m_n = 92 \times 1.007\ 825\ \text{u} + 146 \times 1.008\ 665\ \text{u} = 239.984\ 990\ \text{u}$$

Subtract the mass of ^{238}U from this result:

$$(92 m_p + 146 m_n) - m_{^{238}\text{U}} = 239.984\ 990\ \text{u} - 238.050\ 783\ \text{u} = 1.934\ 207\ \text{u}$$

Multiply the mass difference by c^2 and convert to MeV:

$$E_b = (\Delta m)c^2 = (1.934\ 207\ \text{u})c^2 \left(\frac{931.5\ \text{MeV}/c^2}{\text{u}}\right) = \boxed{1802\ \text{MeV}}$$

and the binding energy per nucleon is $\dfrac{E_b}{A} = \dfrac{1802\ \text{MeV}}{238} = \boxed{7.57\ \text{MeV}}$

19 ••   The neutron, when isolated from an atomic nucleus, decays into a proton, an electron, and an antineutrino as follows: $n \rightarrow {}^1\text{H} + e^- + \bar{\nu}$. The thermal energy of a neutron is of the order of kT, where k is the Boltzmann constant. (a) In both joules and electron volts, calculate the energy of a thermal neutron at 25°C. (b) What is the speed of this thermal neutron? (c) A beam of monoenergetic thermal neutrons is produced at 25°C and has an intensity I. After traveling 1350 km, the beam has an intensity of $\frac{1}{2} I$. Using this information, estimate the half-life of the neutron. Express your answer in minutes.

Picture the Problem The speed of the neutrons can be found from their thermal energy. The time taken to reduce the intensity of the beam by one-half, from I to $I/2$, is the half-life of the neutron. Because the beam is monoenergetic, the neutrons all travel at the same speed.

(a) The thermal energy of the neutron is:

$$E_{thermal} = kT$$
$$= (1.381 \times 10^{-23}\ J/K)(25 + 273)K$$
$$= \boxed{4.11 \times 10^{-21}\ J}$$
$$= 4.11 \times 10^{-21}\ J \times \frac{1\,eV}{1.602 \times 10^{-19}\ J}$$
$$= \boxed{25.7\,meV}$$

(b) Equate $E_{thermal}$ and the kinetic energy of the neutron to obtain:

$$E_{thermal} = \tfrac{1}{2} m_n v^2 \Rightarrow v = \sqrt{\frac{2E_{thermal}}{m_n}}$$

Substitute numerical values and evaluate v:

$$v = \sqrt{\frac{2(4.11 \times 10^{-21}\ J)}{1.673 \times 10^{-27}\ kg}} = \boxed{2.22\,km/s}$$

(c) Relate the half-life, $t_{1/2}$, to the speed of the neutrons in the beam:

$$t_{1/2} = \frac{x}{v}$$

Substitute numerical values and evaluate $t_{1/2}$:

$$t_{1/2} = \frac{1350\,km}{2.22\,km/s} = 608\,s \times \frac{1\,min}{60\,s}$$
$$= \boxed{10.1\,min}$$

21 •• In 1920, 12 years before the discovery of the neutron, Ernest Rutherford argued that proton–electron pairs might exist in the confines of the nucleus in order to explain the mass number, A, being greater than the nuclear charge, Z. He also used this argument to account for the source of beta particles in radioactive decay. Rutherford's scattering experiments in 1910 showed that the nucleus had a diameter of approximately 10 fm. Using this nuclear diameter, the uncertainty principle, and that beta particles have an energy range of 0.02 MeV to 3.40 MeV, show why the hypothetical electrons cannot be confined to a region occupied by the nucleus.

Picture the Problem The Heisenberg uncertainty principle relates the uncertainty in position, Δx, to the uncertainty in momentum, Δp, by $\Delta x \Delta p \geq \frac{1}{2}\hbar$.

Solve the Heisenberg equation for Δp:

$$\Delta p \approx \frac{\hbar}{2\Delta x}$$

Assuming that electrons are contained within the nucleus, the uncertainty in their momenta must be:

$$\Delta p \approx \frac{1.05 \times 10^{-34} \, \text{J} \cdot \text{s}}{2(10 \times 10^{-15} \, \text{m})}$$

$$= 5.25 \times 10^{-21} \, \text{kg} \cdot \text{m/s}$$

The kinetic energy of the electron is given by:

$$K = pc$$

Substitute numerical values and evaluate K for a nuclear electron:

$$K = (5.25 \times 10^{-21} \, \text{kg} \cdot \text{m/s})(2.998 \times 10^{8} \, \text{m/s}) = 1.58 \times 10^{-12} \, \text{J} \times \frac{1 \, \text{eV}}{1.602 \times 10^{-19} \, \text{J}}$$

$$= 9.88 \, \text{MeV}$$

This result contradicts experimental observations that show that the energy of electrons in unstable atoms is of the order of 1 to 1000 eV. Hence the premise on which it is based (that electrons are contained within the nucleus) must be false.

Radioactivity

29 ••   Plutonium is very toxic to the human body. Once it enters the body it collects primarily in the bones, although it also can be found in other organs. Red blood cells are synthesized within the marrow of the bones. The isotope ^{239}Pu is an alpha emitter that has a half-life of 24 360 years. Because alpha particles are an ionizing radiation, the blood-making ability of the marrow is, in time, destroyed by the presence of ^{239}Pu. In addition, many kinds of cancers will also develop in the surrounding tissues because of the ionizing effects of the alpha particles. (*a*) If a person accidentally ingested 2.0 μg of ^{239}Pu and all of it is absorbed by the bones of the person, how many alpha particles are produced per second within the body of the person? (*b*) When, in years, will the activity be 1000 alpha particles per second?

Picture the Problem Each ^{239}Pu nucleus emits an alpha particle whose activity, A, depends on the decay constant of ^{239}Pu and on the number N of nuclei present in the ingested ^{239}Pu. We can find the decay constant from the half-life and the number of nuclei present from the mass ingested and the atomic mass of ^{239}Pu. Finally, we can use the dependence of the activity on time to find the time at which the activity be 1000 alpha particles per second.

(*a*) The activity of the nuclei present in the ingested ^{239}Pu is given by:

$$A = \lambda N \qquad\qquad (1)$$

Find the constant for the decay of ^{239}Pu:

$$\lambda = \frac{\ln(2)}{t_{1/2}} = \frac{0.693}{(24360 \, \text{y})(31.56 \, \text{Ms/y})}$$

$$= 9.02 \times 10^{-13} \, \text{s}^{-1}$$

Express the number of nuclei N present in the mass m_{Pu} of ^{239}Pu ingested:

$$\frac{N}{N_A} = \frac{m_{Pu}}{M_{Pu}} \Rightarrow N = m_{Pu}\frac{N_A}{M_{Pu}}$$

where M_{Pu} is the atomic mass of ^{239}Pu.

Substitute numerical values and evaluate N:

$$N = (2.0\,\mu g)\left(\frac{6.02\times10^{23}\,\text{nuclei/mol}}{239\,\text{g/mol}}\right)$$

$$= 5.04\times10^{15}\,\text{nuclei}$$

Substitute numerical values in equation (1) and evaluate A:

$$A = \left(9.02\times10^{-13}\,\text{s}^{-1}\right)\left(5.04\times10^{15}\,\alpha\right)$$

$$= \boxed{4.55\times10^3\,\alpha/s}$$

(b) The activity varies with time according to:

$$A = A_0 e^{-\lambda t} \Rightarrow t = \frac{\ln\left(\dfrac{A}{A_0}\right)}{-\lambda}$$

Substitute numerical values and evaluate t:

$$t = \frac{\ln\left(\dfrac{1000\,\alpha/s}{4.55\times10^3\,\alpha/s}\right)}{-\left(9.02\times10^{-13}\,\text{s}^{-1}\right)\left(\dfrac{31.56\,\text{Ms}}{y}\right)}$$

$$= \boxed{5.32\times10^4\,y}$$

31 •   The fissile material ^{239}Pu is an alpha emitter. Write the reaction that describes ^{239}Pu undergoing alpha decay. Given that ^{239}Pu, ^{235}U, and an alpha particle have respective masses of 239.052 156 u, 235.043 923 u, and 4.002 603 u, use the relations appearing in Problem 32 to calculate the kinetic energies of the alpha particle and the recoiling daughter nucleus.

Picture the Problem We can write the equation of the decay process by using the fact that the post-decay sum of the Z and A numbers must equal the pre-decay values of the parent nucleus. The Q value in the equations from Problem 32 is given by $Q = -(\Delta m)c^2$.

^{239}Pu undergoes alpha decay according to:

$$\boxed{^{239}_{94}\text{Pu} \rightarrow\, ^{235}_{92}\text{U} + ^4_2\alpha + Q}$$

The Q value for the decay is given by:

$$Q = \left[(m_{Pu}) - (m_U + m_\alpha)\right]c^2\left(\frac{931.5\,\text{MeV}/c^2}{u}\right)$$

Substitute numerical values and evaluate Q:

$$Q = [(239.052156\,u) - (235.043923\,u + 4.002603\,u)]c^2\left(\frac{931.5\text{MeV}/c^2}{u}\right)$$

$$= \boxed{5.24\,\text{MeV}}$$

From Problem 32, the kinetic energy of the alpha particle is given by:
$$K_\alpha = \left(\frac{A-4}{A}\right)Q$$

Substitute numerical values and evaluate K_α:
$$K_\alpha = \left(\frac{239-4}{239}\right)(5.24\,\text{MeV})$$

$$= \boxed{5.15\,\text{MeV}}$$

From Problem 32, the kinetic energy of the ^{235}U is given by:
$$K_U = \frac{4Q}{A}$$

Substitute numerical values and evaluate $K_{^{235}U}$:
$$K_{^{235}U} = \frac{4(5.24\,\text{MeV})}{235} = \boxed{89.2\,\text{keV}}$$

35 •• Radiation has been used for a long time in medical therapy to control the development and growth of cancer cells. Cobalt-60, a gamma emitter that emits photons that have energies of 1.17 MeV and 1.33 MeV, is used to irradiate and destroy deep-rooted cancers. Small needles made of ^{60}Co of a specified activity are encased in gold and used as body implants in tumors for time periods that are related to tumor size, tumor cell reproductive rate, and the activity of the needle. (*a*) A 1.00 μg sample of ^{60}Co, that has a half-life of 5.27 y, that is used to irradiate a small internal tumor with gamma rays, is prepared in the cyclotron of a medical center. Determine the activity of the sample in curies. (*b*) What will the activity of the sample be 1.75 y from now?

Picture the Problem We can use $R_0 = \lambda N$ to find the initial activity of the sample and $R = R_0 e^{-\lambda t}$ to find the activity of the sample after 1.75 y.

(*a*) The initial activity of the sample is the product of the decay constant λ for ^{60}Co and the number of atoms N of ^{60}Co initially present in the sample:
$$R_0 = \lambda N$$

Express N in terms of the mass m of the sample, the molar mass M of ^{60}Co, and Avogadro's number N_A:

$$N = \frac{m}{M} N_A$$

Substituting for N yields:

$$R_0 = \frac{\lambda m N_A}{M}$$

The decay constant is given by:

$$\lambda = \frac{\ln(2)}{t_{1/2}}$$

Substitute for λ to obtain:

$$R_0 = \frac{\ln(2) m N_A}{t_{1/2} M}$$

Substitute numerical values and evaluate R_0:

$$R_0 = \frac{\ln(2)(1.00 \times 10^{-6}\,g)\left(6.022 \times 10^{23}\,\frac{nuclei}{mol}\right)}{\left(5.27\,y \times \frac{31.56\,Ms}{y}\right)\left(60\,\frac{g}{mol}\right)} = 4.183 \times 10^7\,s^{-1} \times \frac{1\,Ci}{3.7 \times 10^{10}\,s^{-1}}$$

$$= \boxed{1.13\,mCi}$$

(b) The activity varies with time according to:

$$R = R_o e^{-\lambda t} = R_o e^{-\left(\frac{0.693 t}{5.27 y}\right)}$$

Evaluate R at $t = 1.75$ y:

$$R = (1.13\,mCi) e^{-\left(\frac{0.693 \times 1.75 y}{5.27 y}\right)}$$

$$= \boxed{0.898\,mCi}$$

Nuclear Reactions

43 •• (a) Use the values 14.003 242 u and 14.003 074 u for the atomic masses of ^{14}C and ^{14}N, respectively, to calculate the Q value (in MeV) for the β-decay reaction $^{14}_{6}C \rightarrow {}^{14}_{7}N + e^- + \bar{v}_e$. (b) Explain why you should not add the mass of the electron to that of atomic ^{14}N for the calculation in (a).

Picture the Problem We can use $Q = -(\Delta m)c^2$ to find the Q values for this reaction.

(a) The masses of the atoms are:

$$m_{^{14}C} = 14.003\ 242\ u$$

$$m_{^{14}N} = 14.003\ 074\ u$$

Calculate the increase in mass:

$$\Delta m = m_f - m_i$$

$$= 14.003\ 074\ u - 14.003\ 242\ u$$

$$= -0.000\ 168\ u$$

Calculate the Q value:

$$Q = -(\Delta m)c^2$$

$$= -(-0.000\ 168\ u)c^2\left(931.5\frac{MeV/c^2}{u}\right)$$

$$= \boxed{0.156\ MeV}$$

(b) The masses given are for atoms, not nuclei, so the atomic masses are too large by the atomic number multiplied by the mass of an electron. For the given nuclear reaction, the mass of the carbon atom is too large by $6m_e$ and the mass of the nitrogen atom is too large by $7m_e$. Subtracting $6me$ from both sides of the reaction equation leaves an extra electron mass on the right. Not including the mass of the beta particle (electron) is mathematically equivalent to explicitly subtracting $1m_e$ from the right side of the equation.

Fission and Fusion

45 •   Assuming an average energy of 200 MeV per fission, calculate the number of fissions per second needed for a 500-MW reactor.

Picture the Problem The power output of the reactor is the product of the number of fissions per second and energy liberated per fission.

Express the required number N of fissions per second in terms of the power output P and the energy released per fission $E_{per\ fission}$:

$$N = \frac{P}{E_{per\ fission}}$$

Substitute numerical values and evaluate N:

$$N = \frac{500\,\text{MW}}{200\,\text{MeV}}$$

$$= \frac{5.00 \times 10^8 \dfrac{\text{J}}{\text{s}} \times \dfrac{1\,\text{eV}}{1.602 \times 10^{-19}\,\text{J}}}{200\,\text{MeV}}$$

$$= \boxed{1.56 \times 10^{19}\,\text{s}^{-1}}$$

47 ••   Consider the following fission reaction:
$^{235}_{92}\text{U} + \text{n} \rightarrow {}^{95}_{42}\text{Mo} + {}^{139}_{57}\text{La} + 2\text{n} + Q$. The masses of the neutron, ^{235}U, ^{95}Mo, and ^{139}La are 1.008 665 u, 235.043 923 u, 94.905 842 u, and 138.906 348 u, respectively. Calculate the Q value, in MeV, for this fission reaction.

Picture the Problem We can use $Q = -(\Delta m)c^2$, where $\Delta m = m_f - m_i$, to calculate the Q value.

The Q value, in MeV, is given by:

$$Q = -(\Delta m)c^2 \left(\frac{931.5\,\text{MeV}/c^2}{\text{u}} \right)$$

Calculate the change in mass Δm:

$$\Delta m = m_f - m_i$$
$$= 94.905\ 842\ \text{u} + 138.906\ 348\ \text{u} + 2(1.008\ 665\ \text{u})$$
$$\qquad\qquad - (235.043\ 923\ \text{u} + 1.008\ 665\ \text{u})$$
$$= -0.223\ 068\ \text{u}$$

Substitute for Δm and evaluate Q:

$$Q = -(-0.223\ 068\ \text{u})c^2 \left(\frac{931.5\,\text{MeV}/c^2}{\text{u}} \right) = \boxed{208\ \text{MeV}}$$

General Problems

53 •   The counting rate from a radioactive source is 6400 counts/s. The half-life of the source is 10 s. Make a plot of the counting rate as a function of time for times up to 1 min. What is the decay constant for this source?

Picture the Problem We can use the given information regarding the half-life of the source to find its decay constant. We can then plot a graph of the counting rate as a function of time.

The decay constant is related to the half-life of the source:

$$\lambda = \frac{\ln(2)}{t_{1/2}} = \frac{\ln(2)}{10\,\text{s}} = 0.0693\,\text{s}^{-1}$$

$$= \boxed{0.069\,\text{s}^{-1}}$$

The activity of the source is given by:

$$R = R_0 e^{-\lambda t} = (6400\,\text{Bq})e^{-(0.0693\,\text{s}^{-1})t}$$

The following graph of $R = (6400\,\text{Bq})e^{-(0.0693\,\text{s}^{-1})t}$ was plotted using a spreadsheet program.

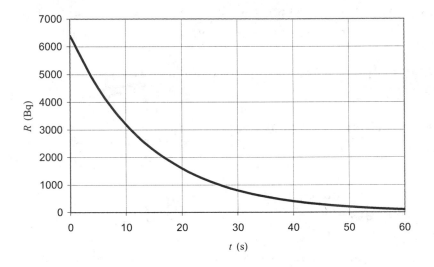

57 •• Show that the ^{109}Ag nucleus is stable and does not undergo alpha decay, $^{109}_{47}\text{Ag} \rightarrow ^{4}_{2}\text{He} + ^{105}_{45}\text{Rh} + Q$. The mass of the ^{109}Ag nucleus is 108.904 756 u, and the products of the decay are 4.002 603 u and 104.905 250 u, respectively.

Picture the Problem We can show that ^{109}Ag is stable against alpha decay by demonstrating that its Q value is negative.

The Q value, in MeV, for this reaction is:

$$Q = -\left[(m_{\text{Rh}} + m_\alpha) - m_{\text{Ag}}\right]c^2\left(931.5\,\frac{\text{MeV}/c^2}{\text{u}}\right)$$

Substitute numerical values and evaluate Q:

$$Q = -\left[(4.002\ 603\ \text{u} + 104.905\ 250\ \text{u}) - 108.904\ 756\ \text{u}\right]c^2\left(931.5\,\frac{\text{MeV}/c^2}{\text{u}}\right)$$

$$= \boxed{-2.88\ \text{MeV}}$$

Remarks: Alpha decay occurs spontaneously and the Q value will equal the sum of the kinetic energies of the alpha particle and the recoiling daughter nucleus, $Q = K_\alpha + K_D$. Kinetic energy cannot be negative; hence, alpha decay cannot occur unless the mass of the parent nucleus is greater than the sum of the masses of the alpha particle and daughter nucleus, $m_P > m_\alpha + m_D$. Alpha decay cannot take place unless the total rest mass decreases.

69 •••   Assume that a neutron decays into a proton and an electron without the emission of a neutrino. The energy shared by the proton and the electron is then 0.782 MeV. In the rest frame of the neutron, the total momentum is zero, so the momentum of the proton must be equal and opposite the momentum of the electron. This determines the ratio of the energies of the two particles, but because the electron is relativistic, the exact calculation of these relative energies is somewhat challenging. (*a*) Assume that the kinetic energy of the electron is 0.782 MeV and calculate the momentum p of the electron in units of MeV/c. *Hint: Use* $E^2 = p^2c^2 + \left(mc^2\right)^2$ *(Equation 39-27).* (*b*) Using your result from Part (*a*), calculate the kinetic energy $p^2/2m_p$ of the proton. (*c*) Because the total energy of the electron and the proton is 0.782 MeV, the calculation in Part (*b*) gives a correction to the assumption that the energy of the electron is 0.782 MeV. What percentage of 0.782 MeV is this correction?

Picture the Problem The momentum of the electron is related to its total energy through $E^2 = p^2c^2 + E_0^2$ and its total relativistic energy E is the sum of its kinetic and rest energies.

(*a*) Relate the total energy of the electron to its momentum and rest energy:

$$E^2 = p^2c^2 + E_0^2 \qquad\qquad (1)$$

The total relativistic energy E of the electron is the sum of its kinetic energy and its rest energy:

$$E = K + E_0$$

Substitute for E in equation (1) to obtain:

$$\left(K + E_0\right)^2 = p^2c^2 + E_0^2$$

Solving for p gives:

$$p = \frac{\sqrt{K\left(K + 2E_0\right)}}{c}$$

Substitute numerical values and evaluate p:

$$p = \frac{\sqrt{(0.782\,\text{MeV})(0.782\,\text{MeV} + 2 \times 0.511\,\text{MeV})}}{c} = \boxed{1.19\,\text{MeV}/c}$$

(b) Because $p_p = -p_e$:

$$K_p = \frac{p_p^2}{2m_p}$$

Substitute numerical values (see Table 7-1 for the rest energy of a proton) and evaluate K_p:

$$K_p = \frac{(1.188\,\text{MeV}/c)^2}{2(938.28\,\text{MeV}/c^2)} = \boxed{752\,\text{eV}}$$

(c) The percent correction is:

$$\frac{K_p}{K} = \frac{752\,\text{eV}}{0.782\,\text{MeV}} = \boxed{0.0962\%}$$

73 •••   Frequently, the daughter nucleus of a radioactive parent nucleus is itself radioactive. Suppose the parent nucleus, designated by P, has a decay constant λ_P; while the daughter nucleus, designated by D, has a decay constant λ_D. The number of daughter nuclei N_D are then given by the solution to the differential equation

$$dN_D/dt = \lambda_P N_P - \lambda_D N_D$$

where N_P is the number of parent nuclei. (a) Justify this differential equation. (b) Show that the solution for this equation is

$$N_D(t) = \frac{N_{P0}\lambda_P}{\lambda_D - \lambda_P}\left(e^{-\lambda_P t} - e^{-\lambda_D t}\right)$$

where N_{P0} is the number of parent nuclei present at $t = 0$ when there are no daughter nuclei. (c) Show that the expression for N_D in Part (b) gives $N_D(t) > 0$ whether $\lambda_P > \lambda_D$ or $\lambda_D > \lambda_P$. (d) Make a plot of $N_P(t)$ and $N_D(t)$ as a function of time when $\tau_D = 3\,\tau_P$, where τ_D and τ_P are the mean lifetimes of the daughter and parent nuclei, respectively.

Picture the Problem We can differentiate $N_D(t) = \frac{N_{P0}\lambda_P}{\lambda_D - \lambda_P}\left(e^{-\lambda_P t} - e^{-\lambda_D t}\right)$ with respect to t to show that it is the solution to the differential equation $dN_D/dt = \lambda_P N_P - \lambda_D N_D$.

(*a*) The rate of change of N_D is the rate of generation of D nuclei minus the rate of decay of D nuclei. The generation rate is equal to the decay rate of P nuclei, which equals $\lambda_P N_P$. The decay rate of D nuclei is $\lambda_D N_D$.

(*b*) We're given that:

$$\frac{dN_D}{dt} = \lambda_P N_P - \lambda_D N_D \tag{1}$$

$$N_D(t) = \frac{N_{P0}\lambda_P}{\lambda_D - \lambda_P}\left(e^{-\lambda_P t} - e^{-\lambda_D t}\right) \tag{2}$$

$$N_P = N_{P0}e^{-\lambda_P t} \tag{3}$$

Differentiate equation (2) with respect to t to obtain:

$$\frac{d}{dt}[N_D(t)] = \frac{N_{P0}\lambda_P}{\lambda_D - \lambda_P}\frac{d}{dt}\left[\left(e^{-\lambda_P t} - e^{-\lambda_D t}\right)\right] = \frac{N_{P0}\lambda_P}{\lambda_D - \lambda_P}\left[-\lambda_P e^{-\lambda_P t} + \lambda_D e^{-\lambda_D t}\right]$$

Substitute this derivative in equation (1) to obtain:

$$\frac{N_{P0}\lambda_P}{\lambda_D - \lambda_P}\left[-\lambda_P e^{-\lambda_P t} + \lambda_D e^{-\lambda_D t}\right] = \lambda_P N_{P0}e^{-\lambda_P t} - \lambda_D\left[\frac{N_{P0}\lambda_P}{\lambda_D - \lambda_P}\left(e^{-\lambda_P t} - e^{-\lambda_D t}\right)\right]$$

Multiply both sides by $\dfrac{\lambda_D - \lambda_P}{\lambda_D \lambda_P}$ and simplify to obtain:

$$\frac{N_{P0}}{\lambda_D}\left[-\lambda_P e^{-\lambda_P t} + \lambda_D e^{-\lambda_D t}\right] = \frac{\lambda_D - \lambda_P}{\lambda_D}N_{P0}e^{-\lambda_P t} - N_{P0}\left(e^{-\lambda_P t} - e^{-\lambda_D t}\right)$$

$$= N_{P0}e^{-\lambda_P t} - \frac{N_{P0}\lambda_P}{\lambda_D}e^{-\lambda_P t} - N_{P0}e^{-\lambda_P t} + N_{P0}e^{-\lambda_D t}$$

$$= -\frac{N_{P0}\lambda_P}{\lambda_D}e^{-\lambda_P t} + N_{P0}e^{-\lambda_D t}$$

$$= \frac{N_{P0}}{\lambda_D}\left[-\lambda_P e^{-\lambda_P t} + \lambda_D e^{-\lambda_D t}\right]$$

which is an identity and confirms that equation (2) is the solution to equation (1).

(*c*) If $\lambda_P > \lambda_D$ the denominator and expression in parentheses are both negative for $t > 0$. If $\lambda_P < \lambda_D$ the denominator and expression in parentheses are both positive for $t > 0$.

(*d*) The following graph was plotted using a spreadsheet program.

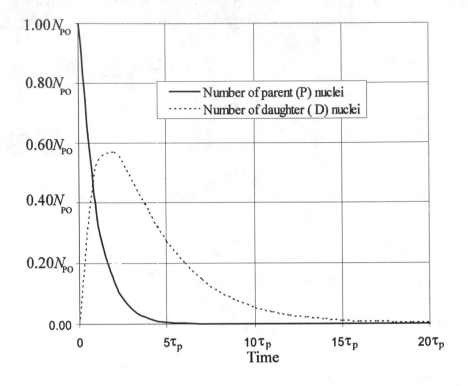

Chapter 41
Elementary Particles and the Beginning of the Universe

Conceptual Problems

3 • How can you tell whether a decay proceeds by the strong interaction or the weak interaction?

Determine the Concept A decay process involving the strong interaction has a very short lifetime ($\sim 10^{-23}$ s), whereas decay processes that proceed by the weak interaction have lifetimes of order 10^{-10} s.

The Conservation Laws

19 •• Using **Error! Reference source not found.** and the laws of conservation of charge number, baryon number, strangeness, and spin, identify the unknown particle, symbolized by (?), in each of the following reactions: (*a*) $p + \pi^- \to \Sigma^0 + (?)$, (*b*) $p + p \to \pi^+ + n + K^+ + (?)$, and (*c*) $p + K^- \to \Xi^- + (?)$

Picture the Problem We can systematically determine Q, B, S, and s for each reaction and then use these values to identify the unknown particles.

(*a*) For the strong reaction:

$$p + \pi^- \to \Sigma^0 + (?)$$

Charge number:

$$+1 - 1 = 0 + Q \Rightarrow Q = 0$$

Baryon number:

$$+1 + 0 = +1 + B \Rightarrow B = 0$$

Strangeness:

$$0 + 0 = -1 + S \Rightarrow S = +1$$

Spin:

$$+\tfrac{1}{2} + 0 = +\tfrac{1}{2} + s \Rightarrow s = 0$$

These properties indicate that the particle is the kaon $\boxed{K^0}$.

(*b*) For the strong reaction:

$$p + p \to \pi^+ + n + K^+ + (?)$$

Charge number:

$$+1 + 1 = +1 + 0 + 1 + Q \Rightarrow Q = 0$$

Baryon number:

$$+1 + 1 = 0 + 1 + 0 + B \Rightarrow B = +1$$

Strangeness:

$$0 + 0 = 0 + 0 + 1 + S \Rightarrow S = -1$$

Spin:

$$+\tfrac{1}{2} + \tfrac{1}{2} = 0 + \tfrac{1}{2} + 0 + s \Rightarrow s = +\tfrac{1}{2}$$

These properties indicate that the particle is either the $\boxed{\Sigma^0}$ or the $\boxed{\Lambda^0}$ baryon.

(c) For the strong reaction:

$$p + \overline{K^-} \rightarrow \Xi^- + (?)$$

Charge number:

$$+1 - 1 = -1 + Q \Rightarrow Q = +1$$

Baryon number:

$$+1 + 0 = +1 + B \Rightarrow B = 0$$

Strangeness:

$$0 - 1 = -2 + S \Rightarrow S = +1$$

Spin:

$$+\tfrac{1}{2} + 0 = +\tfrac{1}{2} + s \Rightarrow s = 0$$

These properties indicate that the particle is the kaon $\boxed{K^+}$.

Quarks

27 ••   Find a possible quark combination for the following particles: (a) Λ^0,
(b) p^-, and (c) Σ^-.

Picture the Problem Because Λ^0, p^-, and Σ^- are baryons, they are made up of three quarks. We can use Tables 41-1 and 41-2 to find combinations of quarks with the correct values for electric charge, baryon number, and strangeness for these particles.

(a) For Λ^0 we need:
$$Q = 0$$
$$B = +1$$
$$S = -1$$

The quark combination that satisfies these conditions is $\boxed{uds.}$

(b) For p^- we need:
$$Q = -1$$
$$B = -1$$
$$S = +1$$

The quark combination that satisfies these conditions is $\boxed{\overline{u}\overline{u}\overline{d}.}$

(c) For Σ^- we need:
$$Q = -1$$
$$B = +1$$
$$S = -1$$

The quark combination that satisfies these condition is $\boxed{dds.}$

The Evolution of the Universe

31 •   A galaxy is receding from Earth at 2.5 percent the speed of light. Estimate the distance from Earth to this galaxy.

Picture the Problem We can use Hubble's law to find the distance from Earth to this galaxy.

The recessional velocity of galaxy is related to its distance by Hubble's law:

$$v - Hr \Rightarrow r - \frac{v}{H}$$

Substitute numerical values and evaluate r:

$$r = \frac{(0.025)c}{\dfrac{23 \text{ km/s}}{10^6 c \cdot y}} = \frac{(0.025)(2.998 \times 10^5 \text{ km/s})}{\dfrac{23 \text{ km/s}}{10^6 c \cdot y}}$$

$$= \boxed{3.3 \times 10^8 \, c \cdot y}$$

General Problems

37 ••   (a) In terms of the quark model, show that the reaction $\pi^0 \rightarrow \gamma + \gamma$ does not violate any conservation laws. (b) Which conservation law is violated by the reaction $\pi^0 \rightarrow \gamma$?

Picture the Problem The π° particle is composed of two quarks, $u\bar{u}$. Hence, the reaction $\pi^\circ \rightarrow \gamma + \gamma$ is equivalent to $u\bar{u} \rightarrow \gamma + \gamma$.

(a) The u and $\bar{u}$ annihilate – resulting in the photons.

(b) Two or more photons are required to conserve linear momentum.